DIGITISING COMMAND AND CONTROL

Human Factors in Defence

Series Editors:

Professor Don Harris, Cranfield University, UK
Professor Neville Stanton, Brunel University, UK
Professor Eduardo Salas, University of Central Florida, USA

Human factors is key to enabling today's armed forces to implement their vision to 'produce battle-winning people and equipment that are fit for the challenge of today, ready for the tasks of tomorrow and capable of building for the future' (source: UK MoD). Modern armed forces fulfil a wider variety of roles than ever before. In addition to defending sovereign territory and prosecuting armed conflicts, military personnel are engaged in homeland defence and in undertaking peacekeeping operations and delivering humanitarian aid right across the world. This requires top class personnel, trained to the highest standards in the use of first class equipment. The military has long recognised that good human factors is essential if these aims are to be achieved.

The defence sector is far and away the largest employer of human factors personnel across the globe and is the largest funder of basic and applied research. Much of this research is applicable to a wide audience, not just the military; this series aims to give readers access to some of this high quality work.

Ashgate's *Human Factors in Defence* series comprises of specially commissioned books from internationally recognised experts in the field. They provide in-depth, authoritative accounts of key human factors issues being addressed by the defence industry across the world.

Digitising Command and Control
A Human Factors and Ergonomics Analysis of Mission Planning and Battlespace Management

NEVILLE A. STANTON
University of Southampton, UK

DANIEL P. JENKINS
Sociotechnic Solutions Ltd, UK

PAUL M. SALMON
Monash University, Australia

GUY H. WALKER
Heriot-Watt University, UK

KIRSTEN M. A. REVELL
University of Southampton, UK

&

LAURA A. RAFFERTY
University of Southampton, UK

CRC Press
Taylor & Francis Group
Boca Raton London New York

CRC Press is an imprint of the
Taylor & Francis Group, an **informa** business

CRC Press
Taylor & Francis Group
6000 Broken Sound Parkway NW, Suite 300
Boca Raton, FL 33487-2742

International Standard Book Number-13: 978-0-7546-7759-8 (Hardback)

Visit the Taylor & Francis Web site at
http://www.taylorandfrancis.com

and the CRC Press Web site at
http://www.crcpress.com

Contents

List of Figures

List of Tables

Acknowledgements

The HFI DTC is a consortium of defence companies and Universities working in cooperation on a series of defence-related projects. The consortium is led by Aerosystems International and comprises Birmingham University, Brunel University, Cranfield University, Lockheed Martin, MBDA and SEA. The consortium was recently awarded The Ergonomics Society President's Medal for work that has made a significant contribution to original research, the development of methodology, and application of knowledge within the field of ergonomics

Aerosystems International	Birmingham University	Brunel University	Cranfield University
Dr Karen Lane	Professor Chris Baber	Professor Neville Stanton	Professor Don Harris
Dr David Morris	Professor Bob Stone	Dr Guy Walker	Andy Farmilo
Linda Wells	Dr Huw Gibson	Dr Daniel Jenkins	Dr Geoff Hone
Kevin Bessell	Dr Robert Houghton	Dr Paul Salmon	Jacob Mulenga
Nicola Gibb	Richard McMaster	Amardeep Aujla	Ian Whitworth
Robin Morrison	Dr James Cross	Kirsten Revell	Dr John Huddlestone
	Robert Guest	Laura Rafferty	Antoinette Caird-Daley

Lockheed Martin UK	MBDA Missile Systems	Systems Engineering and Assessment (SEA) Ltd	University of Southampton
Mick Fuchs	Dr Carol Mason	Dr Anne Bruseberg	Professor Neville Stanton
Lucy Mitchell	Grant Hudson	Dr Iya Solodilova-Whiteley	Dr Guy Walker
Fred Elsey	Chris Vance	Mel Lowe	Kirsten Ravell
Mark Linsell	Steve Harmer	Ben Dawson	Laura Rafferty
Ben Leonard	David Leahy	Jonathan Smalley	Richard McIlroy
Rebecca Stewart		Dr Georgina Fletcher	Linda Sorenson
Jo Partridge			

We are grateful to DSTL who have managed the work of the consortium, in particular (and in alphabetical order) to Geoff Barrett, Bruce Callander, Jen Clemitson, Colin Corbridge, Katherine Cornes, Roland Edwards, Alan Ellis, Helen Forse, Beejal Mistry, Alison Rogers, Jim Squire and Debbie Webb.

This work from the HFI DTC was part-funded by the Human Sciences Domain of the UK Ministry of Defence Scientific Research Programme.

Further information on the work and people that comprise the HFI DTC can be found on www. hfidtc.com.

Glossary

3D	Three dimensional
AH	Abstraction Hierarchy
AoA	Avenue of Approach
BAE	Battlefield Area Evaluation
Bde	Brigade
BG	Battle Group
BS	British Standard
C2	Command and Control
CAST	Command Army Staff Trainer
CCIR	Commander's Critical Information Requirements
CDM	Critical Decision Method
CO	Commanding Officer
CoA	Course of Action
Comms	Communications
CoS	Chief of Staff
COTS	Commercial-off-the-Shelf
CPX	Command Post Exercise
CRI	Colour Rendering Index
CSSO	Combat Service Support for Operations
CWA	Cognitive Work Analysis
dB	decibels
DP	Decision Point
DSA	Distributed Situation Awareness
DSO	Decision Support Overlay
DSOM	Decision Support Overlay Matrix
EEM	External Error Mode
EEMUA	Engineering Equipment & Materials Users Association
EN	European Standardisation
EXCON	Experimental Control Centre
FRAGO	Fragmentary Order
GUI	Graphical User Interface
HCI	Human Computer Interaction
HEI	Human Error Identification
HF	High Frequency
HFI DTC	Human Factors Integration Defence Technology Centre
HQ	Headquarters
HTA	Hierarchical Task Analysis
Hz	Hertz
IR	Information Requests
ISO	International Standards Organisation
ISTAR	Information, Surveillance, Targeting, Acquisition and Reconnaissance
LOP	Local Operational Picture

MoD	Ministry of Defence
MP/BM	Mission Planning and Battlespace Management system
MWL	Mental Work Load
NAI	Named Area of Interest
NATO	North Atlantic Treaty Organisation
NEC	Networked Enabled Capability or Network Centric Warfare
OCOKA	Observation, Cover and Concealment, Obstacles, Key Terrain and Avenues of Approach
OODA	Orientate, Observe, Decide and Act
Op Order	Operational Order
OpsO	Operations Officer
OSPR	Own Situation Position Report
PMV	Predicted Mean Vote
PPD	Predicted Percentage Dissatisfied
RFI	Request For Information
SA	Situation Awareness
SCADA	System Control and Data Acquisition
SHERPA	Systematic Human Error Reduction and Prediction Approach
SME	Subject Matter Expert
SNA	Social Network Analysis
SOI	Standard Operating Instruction
SOP	Standard Operating Procedures
SUDT	Staff User Data Terminal
TAI	Target Area of Interest
UDT	User Data Terminal
UHF	Ultra High Frequency
UK	United Kingdom of Great Britain
USA	United States of America
VDT	Visual Display Terminal
VHF	Very High Frequency
VUDT	Vehicle User Data Terminal
WDA	Work Domain Analysis
WESTT	Workload, Error, Situation awareness, Time and Teamwork method
WO	Warning Order

About the Authors

Professor Neville A. Stanton, Ph.D., F.B.Ps.S., F.Ers.S., M.I.E.T.

HFI DTC, Transportation Research Group, School of Civil Engineering and the Environment, University of Southampton, Highfield, Southampton, SO17 1BJ, UK.
n.stanton@soton.ac.uk

Professor Stanton holds a Chair in Human Factors in the School of Civil Engineering and the Environment at the University of Southampton. He has published over 140 peer-reviewed journal papers and 14 books on Human Factors and Ergonomics. In 1998 he was awarded the Institution of Electrical Engineers Divisional Premium Award for a co-authored paper on Engineering Psychology and System Safety. The Ergonomics Society awarded him the Otto Edholm medal in 2001 and The President's Medal in 2008 for his contribution to basic and applied ergonomics research. In 2007 The Royal Aeronautical Society awarded him the Hodgson Medal and Bronze Award with colleagues for their work on flight deck safety. Professor Stanton is an editor of the journal *Ergonomics* and on the editorial boards of *Theoretical Issues in Ergonomics Science* and the *International Journal of Human Computer Interaction*. Professor Stanton is a Fellow and Chartered Occupational Psychologist registered with The British Psychological Society, and a Fellow of The Ergonomics Society. He has a B.Sc. (Hons) in Occupational Psychology from the University of Hull, a M.Phil. in Applied Psychology and a Ph.D. in Human Factors from Aston University in Birmingham.

Dr Daniel P. Jenkins

Sociotechnic Solutions Ltd, St Albans, Herts, AL1 2LW, UK.

Dr Jenkins graduated in 2004 from Brunel University with an M.Eng. (Hons) in Mechanical Engineering and Design receiving the 'University Prize' for top student in the department. With over 2 years experience as a design engineer in the automotive industry, Dr Jenkins has worked in a number of roles throughout the world. This wide range of placements has developed experience encompassing design, engineering, project management and commercial awareness. He returned to Brunel in 2005 to become a Research Fellow in the Ergonomics Research Group, working primarily on the HFI DTC project. Studying part-time, he gained his Ph.D. in Human Factors and Interaction Design in 2008. Both academically and within industry he has always had a strong focus on customer-orientated design; design for inclusion; and human factors.

Dr Paul M. Salmon

Human Factors Group, Monash University Accident Research Centre, Building 70, Clayton Campus, Monash University, Victoria 3800, Australia.

Dr Salmon is a Senior Research Fellow in the Human Factors Group at Monash University and holds a B.Sc. in Sports Science and an M.Sc. in Applied Ergonomics (both from the University of Sunderland). He has over 6 years experience in applied human factors research in a number of domains, including the military, civil and general aviation, rail and road transport and has previously worked on a variety of research projects in these areas. This has led to him gaining expertise in a broad range of areas, including human error, situation awareness and the application of Human Factors Methods, including human error identification, situation awareness measurement, teamwork assessment, task analysis and cognitive task analysis methods. Dr Salmon's current research interests include the areas of situation awareness in command and control, human error and the application of human factors methods in sport. He has authored and co-authored various scientific journal articles, conference articles, book chapters and books and was recently awarded the Royal Aeronautical Society Hodgson Prize for a co-authored paper in the society's *Aeronautical* journal.

Dr Guy H. Walker

School of the Built Environment, Heriot-Watt University, Edinburgh, EH14 4AS, UK.
Dr Walker read for a B.Sc. Honours degree in Psychology at Southampton University specialising in engineering psychology, statistics and psychophysics. During his undergraduate studies he also undertook work in auditory perception laboratories at Essex University and the Applied Psychology Unit at Cambridge University. After graduating in 1999 he moved to Brunel University, gaining a Ph.D. in Human Factors in 2002. His research focused on driver performance, situational awareness and the role of feedback in vehicles. Since this time he has worked for a human factors consultancy on a project funded by the Rail Safety and Standards Board, examining driver behaviour in relation to warning systems and alarms fitted in train cabs.

Kirsten M. A. Revell

HFI DTC, Transportation Research Group, School of Civil Engineering and the Environment, University of Southampton, Highfield, Southampton, SO17 1BJ, UK.
Ms Revell graduated from Exeter University in 1995 with a B.Sc. (Hons) in Psychology, where her dissertation focused on the use of affordances in product design. After graduating, she spent 6 years working for Microsoft Ltd., implementing and managing the Microsoft Services Academy which prepared graduates for technical and consulting roles across Europe, the Middle East and Africa. In 2005 she undertook a second degree in Industrial Design at Brunel University. As part of her degree, she spent 10 months on industrial placement with the Ergonomics Research Group. During this time, she partook in a major field trial for the HFI DTC, assisting in data collection and analysis. She intends to bring together her psychology and design disciplines by pursuing a Human Factors approach to design, with a particular interest in affordances.

Laura A. Rafferty

HFI DTC, Transportation Research Group, School of Civil Engineering and the Environment, University of Southampton, Highfield, Southampton, SO17 1BJ, UK.
Ms Rafferty completed her undergraduate studies in 2007 graduating with a B.Sc. in Psychology (Hons) from Brunel University. In the course of this degree she completed two industrial placements, the second of which was working as a Research Assistant in the Ergonomics Research Group. During this 7-month period she helped to design, run and analyse a number of empirical studies being run for the HFI DTC. Within this time Ms Rafferty also completed her dissertation exploring

the qualitative and quantitative differences between novices and experts within military command and control. She is currently in her second year of Ph.D. studies, focusing on team work and decision making associated with combat identification.

Preface

This book aims to show how Human Factors and Ergonomics can be used to support system analysis and development. As part of the research work of the Human Factors Integration Defence Technology Centre (HFI DTC), we are often asked to comment on the development of new technologies. For some time now we have looked in-depth at Command and Control activities and functions. The reader is guided to our other books on Modelling Command and Control, Cognitive Work Analysis, Distributed Situation Awareness and Socio-Technical Systems (all published under the *Human Factors in Defence* series) for a fuller appreciation of our work. The research reported in this book brought all of these areas together to look in-depth at a proposal for a new digitised system that would support Command and Control at Brigade Headquarters and below. For us it was a good opportunity to apply the methods we had been developing to a system that was in development. The pages within this book show you how we went about this task and what we found.

It is often the cry of Human Factors and Ergonomics that we are not asked for our involvement in system development early enough. In the past we have written books on Human Factors Methods (published by Ashgate and others), which explain how to apply the methods to system design and evaluation. Here we were given the opportunity, although we also feel that involvement when the system was being tested was too late, as we would have preferred to have been involved in system concept, design and development. Nevertheless, it is pleasing to have been involved in the testing phase, so that any shortcomings could be addressed in subsequent design.

As with all projects of this nature, we have gone to great pains to disguise the system under test for reasons of commercial confidentiality. This means that we are not allowed to disclose the name of the products nor any screen shots of the equipment. We have redrawn all the pictures and removed any reference to the company involved. It is a pity that such steps are required and we wish organisations could be more open about the testing of their products. Any short-term pain would turn into longer-term gain for the products, the users and the organisations involved.

As the contents of this book show, we started our analysis by understanding how mission planning and battlespace management works with traditional materials. The research team not only observed people conducting the tasks but also undertook the training in those tasks themselves. There is much insight to be gained through participant-observation, more than mere observation allows. It also enhanced the understanding of our subsequent observations, because we had performed the tasks for ourselves.

People may approach this book with many different requirements, goals and agenda. For those who want an overview of Human Factors Methods, we recommend chapter two. For those who want to understand mission planning processes, we recommend chapter three. If you are interested in any particular method, read the overview in chapter two, then chapter four for Cognitive Work Analysis, chapter five for Hierarchical Task Analysis, chapter seven for Social Network Analysis, chapter eight for SCADA Analysis, chapter nine for Usability Analysis and chapter ten for Environmental Analysis. For those interested in collaboration and communication in military headquarters, we recommend chapters three, six and seven. Finally, for those interested in our recommendations for future design of digital Command and Control we recommend chapter eleven. We have tried to write each chapter as stand-alone, but accept that people may want to dip in and out of chapters to suit their particular needs. We also feel that this book serves as a perfectly compatible accompaniment to any of our other books on Human Factors Methods, Modelling

Command and Control, Cognitive Work Analysis, Distributed Situation Awareness and Socio-Technical Systems. This book brings all of the topics presented in the previous books together to focus on the analysis of a mission planning and battlespace management system.

Chapter 1
Overview of the Book

This book presents a Human Factors and Ergonomics evaluation of a digital Mission Planning and Battlespace Management (MP/BM) system. Emphasis was given to the activities occurring within Brigade (Bde) and Battle Group (BG) level headquarters (HQ), and the Human Factors team from the HFI DTC distributed their time evenly between these two locations. The insights contained in this volume arise from a wide-ranging and multi-faceted approach, comprising:

- observation of people using the traditional analogue MP/BM processes in the course of their work to understand how analogue MP/BM is used in practice;
- constraint analysis (Cognitive Work Analysis, CWA) of the digital MP/BM system to understand if digital MP/BM is better or worse than the conventional paper-based approach;
- analysis of the tasks and goal structure required by the digital MP/BM, to understand the ease with which the activities can be performed and identify the likely design-induced errors;
- analysis of Distributed Situation Analysis (DSA), to understand the extent to which digital MP/BM supports collaborative working;
- analysis of the social networks that the digital system allows to form spontaneously (to understand the way in which people choose to communicate via voice and data);
- assessment against EEMUA 201 (Engineering Equipment & Materials Users Association) to understand if digital MP/BM meets with best Human Factors practice in control system interface design;
- assessment against a Usability Questionnaire, to gauge user reactions about the ease or difficulty of using the digital MP/BM system); and
- an environmental survey, to understand the extent to which the Bde and BG environment within which people are working meets with British Standard BS/EN/ISO 11064 Environmental Requirements for Control Centres.

A brief summary of the chapters of the book are presented next, with the detailed description of methods, approach, findings and recommendations within the main body of the book.

Chapter two presents an overview of the Human Factors and Ergonomics discipline and the methods associated with it. The discipline is introduced with a few examples of how it has contributed to improved display and control design. This is consistent with the overall aim of improving the well-being of workers, as well as their work, and the general goal of improved system performance. Two examples in particular resonate with the purpose of this book, both taken from aviation over 60 years ago but still with relevance today. Safety of systems is of major importance in Human Factors and safety critical environments have much to gain from its application. Human Factors and Ergonomics offers unique insights into the way in which people work, through the understanding of the interactions between humans, technology, tools, activities, products and their constraints. This understanding is assisted through the application of Human Factors and Ergonomics methods, which are also introduced. Some of these are pursued through the rest of the book. They offer complementary perspectives on the problem and can be used in an integrated manner.

Chapter three presents observational studies of the tasks people were undertaking in the HQs prior to digitisation. The conventional, analogue mission planning process is examined with the objective of identifying the Ergonomics challenges for digitisation. Prototypes of digital mission planning systems are beginning to be devised and demonstrated, but there has been concern expressed over the design of such systems, many of which fail to understand and incorporate the human aspects of socio-technical systems design. Previous research has identified many of the potential pitfalls of failing to take Ergonomic considerations into account, as well as the multiplicity of constraints acting on the planners and planning process. An analysis of mission planning in a BG is presented, focussing on the tasks and the products produced. This study illustrates the efficiency of an analogue process, one that has evolved over many generations to form the Combat Estimate, a process that is mirrored by forces throughout the world. The challenges for digitisation include ensuring that the mission planning process remains easy and involving, preserving the public nature of the products, encouraging the collaboration and cooperation of the planners, and maintaining the flexibility, adaptability and speed of the analogue planning process. It is argued that digitisation should not become an additional constraint on mission planning.

Chapter four presents the constraint analysis performed on the digital MP/BM. This approach, realised through CWA deconstructs the system into different levels of abstraction:

- Functional Purpose (that is, the reason that the system exists; for example, conduct planning to enact higher command's intent);
- Values and Priorities (that is, the measures of success for the system; for example, maintain digital MP/BM effectiveness, minimise casualties, reduce time to generate products and so on);
- Purpose Related Functions (that is, the functions the system is performing; for example, coordination of units, position and status, threat evaluation, resource allocations and so on); and
- Object Related Functions (that is, what the physical objects in the system do; for example, data transmission, voice communication, blue positions, red positions, effects schematic and so on).

This Abstraction Hierarchy (AH) was then used as a basis for interviewing staff officers at BG and Bde level, to find out if digital MP/BM was significantly better, the same, or significantly worse than conventional approaches. The findings showed that the system offered little to support planning, with none of the respondents offering a positive opinion of the system's ability to aid the planning process. Further examination of the results showed that the digital MP/BM estimate process was generally unsupported by the digital system and that in many cases digitisation had a negative effect on tempo, efficiency, effectiveness and flexibility. The participants offered a positive rating for the system's ability to support battlefield management; however, examination of the results reveals that this positive rating can be mainly attributed to the secure voice radio facility rather than the digital MP/BM elements of the system.

Chapter five presents a deconstruction of the activities performed in the operation of the digital MP/BM system. The deconstruction takes place under the rubric of Hierarchical Task Analysis (HTA) and creates a hierarchy of goals, sub-goals and plans. This analysis produced a task model for each of the seven questions in the Combat Estimate. This offers a much higher fidelity of analysis for the steps involved in producing the Combat Estimate products, as overviewed in chapter three. The HTA was used as the basis for investigating the ease or difficultly with which the operations on the digital MP/BM system could be performed. Examples of the difficulties encountered are presented together with suggested remedies in the redesign of the system. The HTA also formed the foundations for human error identification analysis using the Systematic Human Error Reduction and Prediction Approach (SHERPA) method. The SHERPA taxonomy was applied to the HTA in order to identify likely error modes. Examples of the types of design-induced errors that users may be likely to commit are presented. These errors are also intended to focus attention on remedial strategies and stimulate design solutions.

Chapter six presents an evaluation of DSA during mission planning and execution activities supported by the digital MP/BM system. The analysis was undertaken using a mind mapping approach in order to understand how information and knowledge was spread around the various agents (including the human players, artefacts, products and materials). This analysis was split into three parts: Situation Awareness (SA) requirements analysis, analysis of SA during planning tasks and a corresponding analysis for operational execution tasks. The SA requirements analysis indicated that the system is not designed to support the range of distinct user SA requirements present in the MP/BM system. The analysis of the DSA during the planning phases revealed questions about the timeliness and accuracy of information, the tempo of planning in digital MP/BM, the accuracy of information from digital MP/BM (as staff had to engage in additional checking activities) and the poor support for different SA requirements in the different planning cells. Analysis of the operation execution tasks revealed that the Local Operating Picture (LOP) was often out-of-date or spurious (clarification of Own Situation Position Reports (OSPR) data was required, placing more load on the secure voice channel for updates of the blue force positions) and that there was a low level of trust in the LOP and OSPR (requiring the operations cell to compensate for digital MP/BM's shortcomings by drawing blue force positions directly on to the Smartboard – but these were wiped off every time digital MP/BM updated or was changed). In summary, it was found that DSA was not well supported by digital MP/BM as different personnel have different SA requirements, which are subject to change, depending upon their tasks and goals at any moment in time.

Chapter seven considers the analysis of networks in digital Network Enabled Technology. The ideas behind self-synchronisation of people in networks adapting to changes in the environment are presented. Social Network Analysis (SNA) offers the means to populate the NATO SAS-050 model of command and control with live data, ascribing network metrics to each of the NATO model's three primary axes: patterns of interaction (from fully-hierarchical to fully-distributed), distribution of information (from tight-control to broad-control) and decision rights (from unitary to peer-to-peer). This usefully extends the model and enables it to meet several critical needs, firstly, to understand not where a command and control organisation formally places itself in the model but where it 'actually' places itself. Secondly, to see how the command and control organisation's position changes as a result of function and time. And finally, to understand the match between the place(s) the organisation occupies in the so-called 'approach space' and how they map over to a corresponding 'problem space'. In this respect the analysis revealed a mismatch, which powerful examples of emergent user behaviour (in terms of unexpected system use) tried to resolve. In other words, the human-system interaction proved to be highly unstable, but the good news was that the underlying communications architecture was able to facilitate rapid reconfigurations. Using SNA to numerically model the behaviour of the system/organisation provides insight into tempo (with characteristic patterns of reconfigurations in evidence), agility (as modelled by the different areas occupied by the organisation in the NATO model) and self-synchronisation (as evidenced by emergent behaviours). As well as these insights, the modelling work now provides a benchmark for future iterations of the system.

In chapter eight the 'look and feel' of the MP/BM's Human Computer Interface is assessed for compliance with EEMUA 201 guidelines. EEMUA 201 represents accepted industry best practice for the design and operation of control system Human Computer Interfaces. Ideally, the interface for digital MP/BM should be designed to allow staff officers and clerks to conduct their tasks effectively and efficiently. This means in turn that it should conform to their expectations and allow them to find information and perform tasks in a straightforward manner. The EEMUA guidelines are therefore an excellent basis for a review of existing systems or for the design of new systems. EEMUA 201 covers all of the important Human Factors concerns in this setting, such as: the number of screens, navigation techniques, use of windows, screen format and layout considerations. The findings of this analysis show that measured against the 35 EEMUA 201 principles the digital MP/BM only met eight of them. Twelve principles were partially met (for which some improvements to the current system are recommended), whilst a further eight principles failed to be met at all (for which significant shortcomings in design were identified). A further seven of the EEMUA 201 principles were deemed not applicable to digital MP/BM.

Usability assessment was undertaken with a Human Computer Interaction (HCI) questionnaire in chapter nine, which was completed by the staff officers and clerks who used the digital MP/BM at BG (13 respondents) and Bde levels (26 respondents). There were fewer staff in BG, which was reflected in the respondent numbers. The questionnaire comprised nine main sections designed to assess the usability of a particular device or system:

- visual clarity (the clarity with which the system displays information);
- consistency (that is, consistency of the interface in terms of how it looks, the ways in which it presents information and also the ways in which users perform tasks);
- compatibility (that is, the system's compatibility with other related systems);
- informative feedback (that is, the level, clarity and appropriateness of the feedback provided by the system);
- explicitness (that is, the clarity with which the system transmits its functionality, structure and capability);
- appropriate functionality (that is, the level of appropriateness of the system's functionality in relation to the activities that it is used for);
- flexibility and control (that is, the flexibility of the system, and the level of control that the user has over the system);
- error prevention and correction (that is, the extent to which the system prevents user errors from either being made or impacting task performance); and
- user guidance and support (that is, the level of guidance and support that the system provides to its end users).

The overall ratings were generally lower at BG, but even at Bde level the overall ratings failed to go beyond neutral. The system was rated particularly low on 'explicitness' and 'error prevention and correction'. This is mainly because the personnel using the system did not find it intuitive, with some losing work altogether due to inadvertent errors.

For the sake of completeness, chapter ten presents assessments of the physical environment within which digital MP/BM was being used. This was not intended to inform the design of digital MP/BM, rather it was to consider if the surrounding environment met with current standards for control centres (that is, BS/EN/ISO 11064 Environmental Requirements for Control Centres). Whilst the comfortable and benign operational environment found in civilian domains, and to which current best practice and guidelines applies, may not be directly relevant to military domains, there remains an inviolable duty of care and opportunities to learn lessons. From a human performance point of view, the command and control environment is generally too cold. Digitisation, and the requirement this brings for sedentary computer-based work, is unlikely to improve this situation. Noise levels approach harmful long-term exposure levels and maintained levels are well in excess of best practice. Air quality (and associated low-level health symptoms) is poor. Lighting would also fail to meet comparable civilian standards; it is too dark overall but has poor directivity meaning that, paradoxically, there is also too much glare. Given the safety critical nature of the tasks being undertaken, comparison against acknowledged best practice sees the present environment as being sub-optimal for safe and efficient human performance.

Chapter eleven of the book presents a summary of all the preceding chapters, drawing all of the main findings together. Conclusions on the extent to which the digital MP/BM system meets Human Factors and Ergonomics criteria are drawn from the Constrains Analysis, HTA, DSA, SNA, System Control and Data Acquistion (SCADA) Analysis, Usability Analysis and Environmental Survey. This represents a very thorough assessment of any new system, but digitisation brings with it additional requirements that have important ramifications, and therefore cannot be undertaken lightly. The recommendations for short-term improvements in the current generation of digital MP/BM system are divided into five sections: general design improvements, user-interface design improvements, hardware improvements, infrastructure improvements and support improvements. Looking forward to next generation digital MP/BM systems, general Human Factors design principles are presented and Human Factors issues in digitising mission planning are considered. Future system design would do well to consider the Human Factors methods presented in chapter two, the understanding gained from mission planning demands and constraints in chapter three, together with the insights gained in the various analysis from chapters four to ten. The design of the digital MP/BM system should not become one of the operational constraints in mission planning and battlespace management. The science and practice of Human Factors has much to offer in terms of resolving this situation, provided that it is applied at the beginning (and then throughout) the system design lifecycle, rather than at the end when its impact is significantly diminished.

Chapter 2
Human Factors in System Design

Human Factors

Human Factors and Ergonomics have over 100 years history in the UK and USA, from humble beginnings at the turn of the last century to the current day. A detailed account of the historical developments in both the USA and UK may be found in Meister (1999). This account even covers the pre-history of the discipline. To cut a long story short, the discipline emerged out of the recognition that analysis of the interaction between people and their working environment revealed how work could be designed to reduce errors, improve performance, improve quality of work and increase the work satisfaction of the workers themselves. Two figures stand out at the early beginnings of the discipline in the 1900s, Frank and Lillian Gilbreth (Stanton, 2006). The Gilbreths sought to discover more efficient ways to perform tasks. By way of a famous example of their work, they observed that bricklayers tended to use different methods of working. With the aim of seeking the best way to perform the task, they developed innovative tools, job aids and work procedures. The resultant effect of these changes to the work meant that the laying of a brick had been reduced dramatically from approximately 18 movements by the bricklayer down to some four movements. Thus the task was therefore performed much more efficiently. This analysis amongst others led to the discovery of 'laws of work', or 'Ergo-nomics' as it was called (Oborne, 1982). Although the discipline has become much more sophisticated in the way it analyses work (as indicated in the next section), the general aims to improve system performance and quality of working life remain the principle goals. Human Factors and Ergonomics has been defined variously as 'the scientific study of the relationship between man and his working environment' (Murrell, 1965), 'a study of man's behaviour in relation to his work' (Grandjean, 1980), 'the study of how humans accomplish work-related tasks in the context of human-machine systems' (Meister, 1989), 'applied information about human behaviour, abilities, limitations and other characteristics to the design of tools, machines, tasks, jobs and environments' (Sanders & McCormick, 1993), and 'that branch of science which seeks to turn human-machine antagonism into human-machine synergy' (Hancock, 1997). From these definitions, it may be gathered that the discipline of Human Factors and Ergonomics is concerned with: human capabilities and limitations, human-machine interaction, teamwork, tools, machines and material design, environments, work and organisational design. The definitions also place some implied emphasis on system performance, efficiency, effectiveness, safety and well-being. These remain important aims for the discipline.

The role Ergonomics has to play in the design of displays of information and input controls is particularly pertinent to the contents of this book, as the main focus is on the design of digital Mission Planning and Battlespace Management (MP/BM) systems. The genus of Ergonomics in military systems display and control design can be traced back to the pioneering works of Paul Fitts and Alphonse Chapanis in aviation. Chapanis (1999) recalls his work at the Aero Medical Laboratory in the early 1940s where he was investigating the problem of pilots and co-pilots retracting the landing gear instead of the landing flaps after landing. His investigations in the B-17 (known as the 'Flying Fortress') revealed that the toggle switches for the landing gear and the landing flaps were both identical and next to each other. Chapanis's insight into human performance enabled him to understand how the pilot might have confused the two toggle switches, particularly after the stresses of a combat mission.

He proposed coding solutions to the problem: separating the switches (spatial coding) and/or shaping the switches to represent the part they control (shape coding), so the landing flap switch resembles a 'flap' and the landing gear switch resembles a 'wheel'. Thus the pilot can tell by looking at, or touching, the switch what function it controls. In his book, Chapanis also proposed that the landing gear switch could be deactivated if sensors on the landing struts detected the weight of the aircraft.

Grether (1949) reports on the difficulties of reading the traditional three-needle altimeter which displays the height of the aircraft in three ranges: the longest needle indicates 100s of feet, the broad pointer indicates 1,000s of feet and the small pointer indicates 10,000s of feet. The work of Paul Fitts and colleagues had previously shown that pilots frequently misread the altimeter. This error had been attributed to numerous fatal and non-fatal accidents. Grether devised an experiment to see if different designs of altimeter could have an effect on the interpretation time and the error rate. If misreading altimeters was really was a case of 'designer error' rather than 'pilot error' then different designs should reveal different error rates. Grether tested six different variations of the dial and needle altimeter containing combinations of three, two and one needles with and without an inset counter as well as three types of vertically moving scale (similar to a digital display). Pilots were asked to record the altimeter reading. The results of the experiment showed that there were marked differences in the error rates for the different designs of the altimeters. The data also show that those displays that took longer to interpret also produced more errors. The traditional three-needle altimeter took some 7 seconds to interpret and produced over 11 per cent errors of 1,000 feet or more. By way of contrast, the vertically moving scale altimeters took less than 2 seconds to interpret and produced less than 1 per cent errors of 1,000 feet or more.

Both of these examples, one from control design and one from display design, suggest that it is not 'pilot error' that causes accidents; rather it is 'designer error'. This notion of putting the blame on the last person in the accident chain (for example, the pilot), has lost credibility in modern Ergonomics. Modern day researchers take a systems view of error, by understanding the relationships between all the moving parts in a system, both human and technical, from concept, to design, to manufacture, to operation and maintenance (including system mid-life upgrades) and finally to dismantling and disposal of the system. These stages map nicely onto the UK MoD's CADMID life cycle stages (Concept – Assessment – Design – Manufacture – In-service – Disposal).

The term 'Human Factors' seems to have come from the USA to encompass any aspect of system design, operation, maintenance and disposal that has bearing on input or output. The terms Human Factors and Ergonomics are often used interchangeably or together. In the UK, Ergonomics is mostly used to describe physiological, physical, behavioural and environmental aspects of human performance whereas Human Factors is mostly used to describe cognitive, social and organisational aspects of human performance. Human Factors and Ergonomics encompass a wide range of topics in system design, including: Manpower, Personnel, Training, Communications Media, Procedures, Team Structure, Task Design, Allocation of Function, Workload Assessment, Equipment Design, System Safety and Health Hazards. The term Human Factors will be used throughout this book, although this may also mean Ergonomics. Modern day Human Factors focuses on integration with other aspects of System Engineering. According to the UK MoD, Human Factors Integration is about '... *providing a balanced development of both the technical and human aspects of equipment provision. It provides a process that ensures the application of scientific knowledge about human characteristics through the specification, design and evaluation of systems.* ' (MoD, 2000, p. 6). This book focuses on the examination of a digital command and control system that was developed for both mission planning and battlespace management. Human Factors methods have been developed over the past century, to help design and evaluate new systems. These methods are considered in the following section.

Human Factors Methods

Human Factors methods are designed to improve product design by understanding or predicting user interaction with the devices (Stanton & Young, 1999); these approaches have a long tradition in system design and tend to have greater impact (as well as reduced cost) when applied early on in the design process (Stanton & Young, 1999), long before the hard-coding and hard-build has begun. The design life cycle has at least ten identifiable stages from the identification of product need up to product release, namely: identification of design need, understanding the potential context of product use, development of concepts, presentation of mock-ups, refinement of concepts, start of coding and hard-build, iterative design process, release of a prototype, minor refinements and then release of the first version of the product. As illustrated in Figure 2.1 by the brown shaded area, there is often far too little Human Factors effort involved far too late in the design process (the dark shaded area is a caricature of the Human Factors effort involved in developing the digital MP/BM system). Ideally, the pattern should be reversed, with the effort front-loaded in the project (as illustrated by the light shaded area). Such a strategy would undoubtedly have led to a much improved digital MP/BM system, at considerably reduced cost and with more timely delivery.

As a rough heuristic, the more complex a product is the more important Human Factors input becomes. Complexity is not a binary state of either 'complex' or 'not complex', the level of complexity lies on a non-numerical scale that can be defined through a set of heuristics (Woods, 1988):

- Dynamism of the system: To what extent can the system change states without intervention from the user? To what extent can the nature of the problem change over time? To what extent can multiple ongoing tasks have different time spans?
- Parts, variables and their interconnections: The number of parts and the extensiveness of interconnections between the parts or variables. To what extent can a given problem be due to multiple potential causes and to what extent can it have multiple potential consequences?
- Uncertainty: To what extent can the data about the system be erroneous, incomplete or ambiguous – how predictable are future states?
- Risk: What is at stake? How serious are consequences of users' decisions?

The environment that the command and control system operates within can be seen to be highly complex. The system is dynamic as it frequently changes in unexpected ways, there are a huge number of interconnected parts within the system, data within the system is frequently incorrect or out of date, and the risk inherent in the system is 'life or death'.

It is to the British Army's immense credit that even the most recalcitrant of equipment issues 'can be made to work', but the new era of networked interoperability (and the complexity this brings) challenges even this ability. Whilst convoluted 'workarounds' may appear to overcome some of the system design problems in the short term, opportunities may be lost in gaining greater tempo, efficiency, effectiveness, flexibility and error reduction. Indeed, this practice is, arguably, fast becoming an optimum strategy for increasing error potential, reducing tempo, efficiency and operational effectiveness. A new era of networked interoperability requires a new approach, one that confronts the challenges of harnessing human capability effectively using structured methodologies.

There are a wide range of Human Factors methods available for the analysis, design and evaluation of products and systems. A detailed description of the most commonly used approaches can be found in Stanton et al. (2005a, b). The choice of method used is influenced by a number of factors; one of these factors is the stage in the design process. By way of an example of how

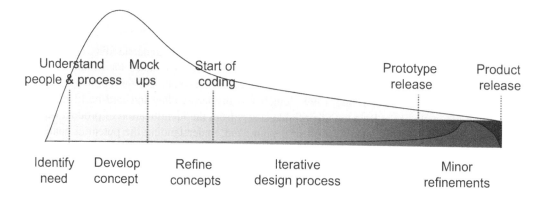

Understand   Mock   Start of   Prototype   Product
people & process ups   coding   release   release

Identify   Develop   Refine   Iterative   Minor
need   concept   concepts   design process   refinements

Figure 2.1   Illustration showing Human Factors effort is better placed in early stages of the design process

structured approaches to human/system integration can be employed, Figure 2.2 relates a number of specific methods to the design life cycle.

Figure 2.2 shows at least 11 different types of Human Factors methods and approaches that can be used through the design life cycle of a new system or product (Stanton & Young, 1999; Stanton et al., 2005a, 2005b). As the figure shows, many of these are best applied before the software coding and hard-build of a system starts. The approach places emphasis on the analysis and development of early prototypes. The assessment described within this book has come very late in the design process, far too late to make fundamental changes in system design; even small design modifications would be costly at this stage. It is extremely likely that an early involvement of Human Factors expertise and methodology would have resulted in a better implementation of the digital MP/BM system. The Human Factors methods advocated within the design life cycle are described further below. The methods starred (thus*) were applied to the case study presented within this book.

Cognitive Work Analysis (CWA)* is a structured framework for considering the development and analysis of complex socio-technical systems. The framework leads the analyst to consider the environment the task takes place within and the effect of the imposed constraints on the system's ability to perform its purpose. The framework guides the analyst through the process of answering the questions of why the system exists and what activities are conducted within the domain, as well as how this activity is achieved and who is performing it. The analysis of constraints provides the basic formulation for development of the early concept for the system and the likely activities of the actors within it. Thus CWA offers a formative design approach.

Systems design methods are often used to provide structure to the design process, and also to ensure that the end-user of the product or system in question is considered throughout the design process. For example, allocation of function analysis is used by system designers to determine whether jobs, tasks and system functions are allocated to human or technological agents within a particular system. Focus group approaches use group interviews to discuss and assess user opinions and perceptions of a particular design concept. In the design process, design concepts are evaluated by the focus group and new design solutions are offered. Scenario-based design involves the use of scenarios or storyboard presentations to communicate or evaluate design concepts. A set of scenarios depicting the future use of the design concept are proposed and performed, and the design concept is evaluated. Scenarios typically use how, why and what-if questions to evaluate and modify a design concept.

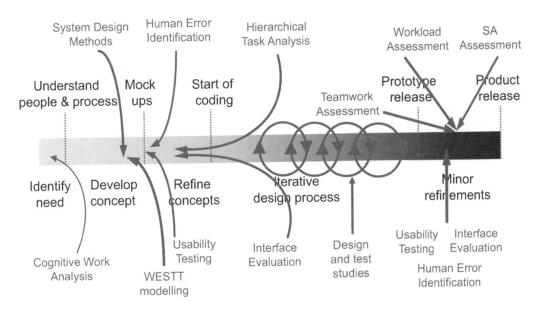

Figure 2.2   Application of Human Factors methods by phase of the design process

Workload, Error, Situation Awareness, Time and Teamwork (WESTT) is a Human Factors tool produced under the aegis of the HFI DTC. The aim of the tool is to integrate a range of Human Factors analyses around a tripartite closely-coupled network structure. The three networks are of Task, Knowledge and Social networks, and are analysed to identify their likely effects on system performance. The WESTT tool developed by the HFI DTC models potential system performance and therefore enables the analyst to consider alternative system structures.

Usability testing* methods are used to consider the usability of software on three main dimensions from ISO9241-11: effectiveness (how well does the product performance meet the tasks for which it was designed?); efficiency (how much resource, for example, time or effort, is required to use to the product to perform these tasks?) and attitude (for example, how favourably do users respond to the product?). It is important to note that it is often necessary to conduct separate evaluations for each dimension rather than using one method and hoping that it can capture all aspects.

Human Error Identification (HEI)* methods can be used either during the design process to highlight potential design induced error, or to evaluate error potential in existing systems. HEI works on the premise that an understanding of an employee's work task and the characteristics of the technology being used allows us to indicate potential errors that may arise from the resulting interaction (Baber and Stanton, 1996). The output of HEI techniques usually describes potential errors, their consequences, recovery potential, probability, criticality and offers associated design remedies or error reduction strategies. HEI approaches can be broadly categorised into two groups, qualitative and quantitative techniques. Qualitative approaches are used to determine the nature of errors that might occur within a particular system, whilst quantitative approaches are used to provide a numerical probability of error occurrence within a particular system. There is a broad range of HEI approaches available to the HEI practitioner, ranging from simplistic External Error Mode (EEM) taxonomy-based approaches to more sophisticated human performance simulation techniques.

Hierarchical Task Analysis (HTA)* is used to describe systems in terms of their goals and sub-goals. HTA works by decomposing activities into a hierarchy of goals, subordinate goals, operations and plans, which allows systems to be described exhaustively. There are at least 12 additional applications to which HTA has been put, including interface design and evaluation, training, allocation of functions, job description, work organisation, manual design, job aid design, error prediction and analysis, team task analysis, workload assessment and procedure design. These extensions make HTA particularly useful in system development when the design has begun to crystallise.

Interface Evaluation* methods are used to assess the human-machine interface of a particular system, product or device. These methods can be used to assess a number of different aspects associated with a particular interface, including user performance, user satisfaction, error, layout, labelling, and the controls and displays used. The output of interface analysis methods is then typically used to improve the interface through redesign. Such techniques are used to enhance design performance, through improving the device or system's usability, user satisfaction, and reducing user errors and interaction time.

Design and Test studies are needed to determine if any measured differences between the new systems and their baselines are real, statistically significant, differences that are likely to generalise beyond the cases studied. There are two broad approaches for Design and Test studies: quantitative and qualitative. Quantitative testing is a formal, objective, systematic process in which numerical data is utilised to obtain information. Quantitative testing tends to produce data that compare one design over another or data that compare a design against a benchmark. Qualitative testing considers opinions and attitudes toward designs. Whilst the attitudes can be measured on scales, often the approach involves an in-depth understanding of the reasons underlying human behaviour. Whilst quantitative studies are concerned with relative differences in performance, qualitative studies are concerned with the reasons for those differences. Typically quantitative research requires larger random samples whereas qualitative research requires smaller purposely selected samples.

Teamwork Assessment* methods are used to analyse those instances where actors within a team or network coordinate their behaviour in order to achieve tasks related to the team's goals. Team-based activity involves multiple actors with multiple goals performing both teamwork and task-work activity. The activity is typically complex (hence the requirement for a team) and may be dispersed across a number of different geographical locations. Consequently there are a number of different team performance techniques available to the Human Factors practitioner, each designed to assess certain aspects of team performance in complex systems. The team performance techniques can be broadly classified into the following categories: team task analysis techniques; team cognitive task analysis techniques; team communication assessment techniques; team Situation Awareness (SA) measurement techniques; team behavioural assessment techniques; and team Mental Work Load (MWL) assessment techniques.

Workload Assessment should be used throughout the design life cycle, to inform system and task design as well as to provide an evaluation of workload imposed by existing operational systems and procedures. There are a number of different workload assessment procedures available. Traditionally, using a single approach to measure workload has proved inadequate, and as a result a combination of the methods available is typically used. The assessment of workload may require a battery of techniques, including primary task performance measures, secondary task performance measures (reaction times, embedded tasks), physiological measures and subjective rating techniques.

Situation Awareness (SA)* refers to an individual's, team's or system's awareness of 'what is going on' (Endsley, 1995a). SA measures are used to measure level of awareness during task performance. The assessment of SA can be used throughout the design life cycle, either to determine the levels of SA provided by novel technology or design, or to assess SA in existing operational systems. SA measures are necessary in order to evaluate the effect of new technologies and training interventions upon SA, to examine factors that affect SA, to evaluate the effectiveness of processes and strategies for acquiring SA and in investigating the nature of SA itself. There are a number of different SA assessment approaches available; in a review of SA measurement techniques, Salmon et al. (2006) describe various approaches, including performance measures, external task measures, embedded task measures, subjective rating techniques (self and observer rating), questionnaires (post-trial and on-line) and the freeze technique.

An integrated design approach is advocated, where the project team includes Human Factors throughout the design life cycle, which will be able to translate user requirements into design specification. As indicated by the stars (thus*) a variety of methods were used in the case study presented in the following chapters. The Human Factors approach also advocates an 'interface first' approach, within a rapid prototyping, development and testing cycle. As the interface concepts crystallise, then hard-coding and hard-build may begin. Details of all the methods discussed may be found in Stanton et al. (2005a, 2005b). This should be the starting point for any future work.

Chapter 3
Mission Planning and Battlespace Management

The Planning Process at Battle Group

Mission failure is often thought to be the result of poor mission planning (Levchuk et al., 2002), which places considerable demands on the planners and the planning process. This observation is further confounded by the two general principles of warfare. The first principle is that of the 'fog of war' (that is, the many uncertainties about the true nature of the environment – Clausewitz, 1832) and second the principle that 'no battle plan survives contact with the enemy' (that is, no matter how thorough the planning is, the enemy is unlikely to be compliant and may act in unpredictable ways – von Moltke, undated). These three tenets (that is, the effects of uncertainty, the enemy and failure on mission planning) require the planning process to be robust, auditable and flexible. Mission planning has to be a continuous, iterative and adaptable process, optimising mission goals, resources and constraints (Levchuck, 2002). Roth et al. (2006) argue that the defining characteristic of command and control is the continual adaptation to a changing environment. Constant change in the goals, priorities, scale of operations, information sources and systems being used means that the planning systems need to be extremely adaptable to cope with these changes. According to Klein and Miller (1999) there are many constraints acting on mission planning, including scarcity of resources, time pressure, uncertainty of information, availability of expertise and the structure of the tasks to be undertaken. Mission planning requires knowledge of the domain, objects in the domain and their relationships as well as the constraints acting on the domain, the objects and their relations (Kieweit et al., 2005). Klein and Miller (1999) also note that the planning cycles can range from a couple of hours to a few days depending upon the complexity of the situation and the time available. Given all of the constraints acting on the planning process and the need for the plan to be continually revised and modified in the light of the enemy actions and changing situation, Klein and Miller (1999) argue that 'simpler plans might allow better implementation and easier modification' (p. 219). This point is reinforced by Riley et al. (2006) who assert that 'plans need to be simple, modifiable, flexible, and developed so that they are quickly and easily understood' (p. 1143).

Mission planning is an essential and integral part of battle management. Although there are some differences within and between the armed services (and the coalition forces) in the way they go about mission planning, there are also some generally accepted aspects that all plans need to assess. These invariants include: enemy strength, activity and assumed intentions, the goals of the mission, analysis of the constraints in the environment, the intent of the commander, developing Courses of Action (CoAs), choosing a CoA, identifying resources requirements, synchronising the assets and actions, and identifying control measures. A summary of the planning process for the US land military may be found in Riley et al. (2006) and the Canadian land forces may be found in Prefontaine (2002). Their description has much in common with land-based planning in the British Army, which is described in The Combat Estimate booklet (CAST, 2007).

The mission planning process has been observed by the authors at the Land Warfare Centre at Warminster and on training exercises in Germany. The observations at Warminster have been both as participant-observers and as normal observers. This section describes the observed activities in the planning process following a Warning Order (WO) received from Brigade (Bde). For the purpose of this analysis, only the conventional materials (whiteboards, maps, overlays, paper, flipcharts and staff officers' notebooks) will be examined. As Figure 3.1 shows, the planning is undertaken in a 'public' environment when various people contribute and all can view the products. This 'public' nature of the products is particularly useful at the briefings, which encourages collaboration and cooperation. It also helps to focus the planners' minds on the important issues and the command intent.

Warning Order from Brigade arrived

The WO arrived and was handed to the Chief of Staff (CoS) who read the whole document. The CoS read the whole document first, highlighting relevant material for themself and the Company level.

Chief of Staff Creates Company Warning Order

The WO was too detailed for Company level, so some editing by CoS was necessary, as well as the inclusion of some additional material to clarify the anticipated task requirements.

Figure 3.1   Battle Group Headquarters

Send Warning Order to Companies

The modified and edited WO was then sent to the companies below the Battle Group (BG), so that they would have advance notice of the intention of the orders when they arrived. This gives them an opportunity to prepare in advance of the actual orders.

Create Planning Timeline

The CoS created a planning timeline for the production of a plan to defeat an assault team that had parachuted into their area. There were 2 hours available to construct the plan (from 13:00 to 15:00), which meant approximately 17 minutes per question (of the Combat Estimate's seven questions). The planning timeline was drawn on a flipchart as shown in Figure 3.2.

Question 1. What is the Enemy Doing and Why?

Question 1 was undertaken by the Engineer and the Intelligence Officer in parallel with question 2. Key terrain features were marked (such as slow-go areas like forests and rivers), as were the approximate disposition of the enemy forces and likely locations, potential Avenues of Approach (AoAs), and likely CoA (see Figure 3.3). In this case, it was thought that the enemy assault force was likely to try and meet up with the main armoured forces approaching from the west. The

Planning time-line

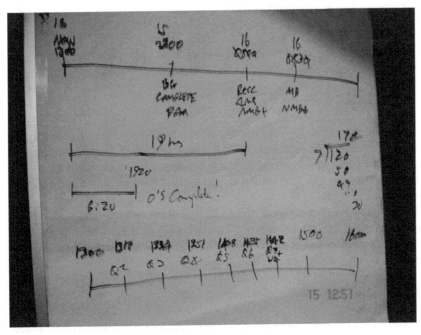

Battle Group level plan

Two hours planning time

120/7 = 17 minutes per question

Figure 3.2   Planning timeline on a flipchart

What is the enemy doing?

Enemy location and strength

Mobility corridors

Go and slow-go areas

Known concepts and doctrine

Figure 3.3   Threat integration on map and overlay

enemy had landed in an area surrounded by forest which gave them some protection, although it was thought that they had not landed where they intended.

Question 2. What have I been Told to do and Why?

The CoS interpreted the orders from Bde together with the BG Commander to complete the Mission Analysis. Each line of the orders was read and the specified and implied tasks were deduced. These were written by hand on to a whiteboard as shown in Figure 3.4. The Commander's Critical Information Requirements (CCIRs) and Information Requests (IRs) were identified and noted for each task, when appropriate.

When the CCIRs/IRs had been completed, the CoS read them off the Mission Analysis whiteboard (expanding where necessary to improve intelligibility) to a clerk who typed them directly on to the Requests For Information (RFI) sheet. The requests were radioed up to Bde and the responses were tracked on the whiteboard.

Question 3. What Effects do I Want to Have on the Enemy?

The CO then drew their required effects on to a flipchart (see Figure 3.5). Three effects were placed above the planning line (SCREEN, CLEAR and DEFEAT) and four effects were placed below the planning line (SCREEN, DEFEAT, GUARD and DEFEND). The two SCREEN effects were placed to prevent the enemy from the west coming to the aid of the group who were being attacked.

What have I been told what to do?

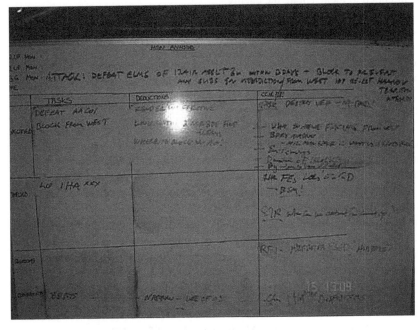

Interpret orders

Identify explicit tasks

Deduce implicit tasks

Raise CCIRs and RFIs

Figure 3.4   Mission Analysis on a whiteboard

What effects do I want to have?

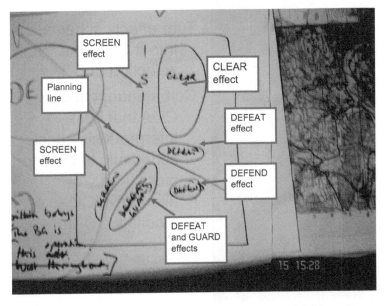

Understand Threat and mission

Command direction

Main effects

Rough outline

Figure 3.5   Effects Schematic drawn on a flipchart and laid on the map

The CLEAR effect was intended to remove any enemy from the forest, if they were there. The DEFEAT effect was intended to incapacitate the enemy.

Question 4. Where can I Best Accomplish Each Action/Effect?

The CoS and BG Commander worked on three CoAs to achieve the Commander's effects, as shown in Figure 3.6. This was a very quick way to propose and compare three potential CoAs in response to the CO's Effects Schematic (remembering that the planning timeline only allowed 17 minutes for each of the seven questions of the Combat Estimate).

Meanwhile the Engineer took the CO's Effects Schematic and put the effects onto the ground, using a TALC (TALC is believed to come from Talc Mica [a crystalline mineral which can be used as a glass substitute], in the military it refers to a clear plastic sheet on which map overlays are drawn) on a paper map (see Figure 3.7). Each effect became either a Named Area of Interest (NAI) or a Target Area of Interest (TAI). Decision Points (DP) were placed between NAIs and TAIs. The resultant overlay is called the Decision Support Overlay (DSO).

It is worth noting that it took approximately 15 minutes to construct the DSO on the TALC (by the Engineer). Between the end of question four (Where can I best accomplish each action and effect?) and the start of question five (What resources do I need to accomplish each action and effect?) the Combat Estimate process was interrupted by the return of the CO, who requested a briefing. The CO reviewed the CoAs and made some revisions, which was followed by a briefing by the CoS to all cells in the HQ.

Where can I best accomplish each effect?

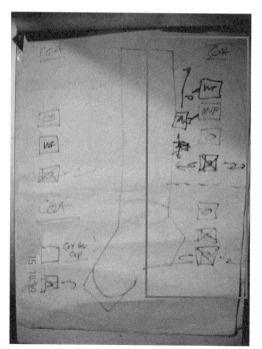

Turn effects schematic into a COA or a selection of COAs

Commander chooses COA that best interprets his effects schematic or cherry picks from alternative COAs or modifies COA(s)

Figure 3.6   COAs developed on a flipchart

Where can I best accomplish each effect?

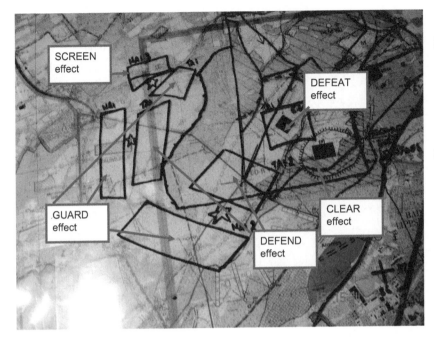

Chosen COA mapped onto map

Check relation between NAI, DP and TAI

Figure 3.7   DSO on map and overlay

Question 5. What Resources do I Need to Accomplish Each Action/Effect?

The Engineer then constructed the Decision Support Overlay Matrix (DSOM) on paper, taking the NAIs, TAIs and DPs from the paper map and linking them to each other, their location and purpose, and the asset that would be used to achieve the effect (see Figure 3.8). There is a clear link between the NAIs, TAIs and on the hand-written flipchart. The manual production of the DSOM on the paper flipchart offers a process of checking the logic of the DSO, making sure that the NAIs, TAIs and DPs link together and that the assets are being deployed to best effect (that is, relating each asset to a purpose as the columns are next to each other in the flipchart version of the DSOM).

Question 6. When and Where do the Actions Take Place in Relation to Each Other?

The CoS led the discussion of how the force elements would move together through the battle (through a mixture of forward recce, mounted and dismounted troops, and armoured vehicles) with logistical support and coordinated indirect fire ahead of them (controlled by the fire control lines – see question seven). This was enacted on the map from the start position (on the left of Figure 3.9) to the end position (on the right of Figure 3.9) to capture the synchronisation issues, which were recorded on to the Coordination Measures whiteboard (see Figure 3.10).

The coordination measures were used as a precursor to the construction of the synchronisation matrix.

What resources do I need?

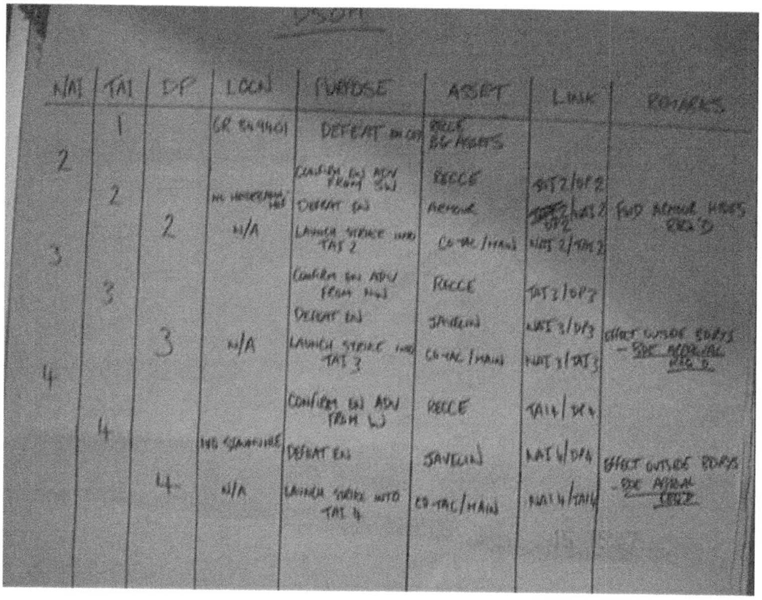

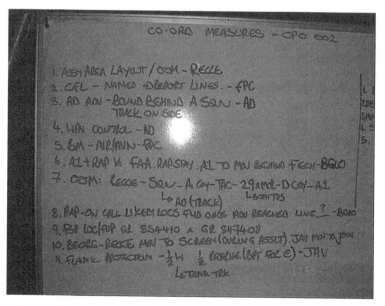

Map relations between NAI, TAI and DP

Identify location, purpose, and asset.

Figure 3.8   DSOM on a flipchart

Where and when to actions take place?

Synchronise actions and co-ordinate assets

Figure 3.9   Coordination of force elements on map and overlay via a wargame

Where and when to actions take place?

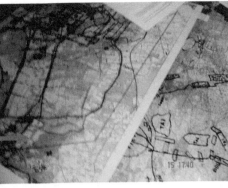

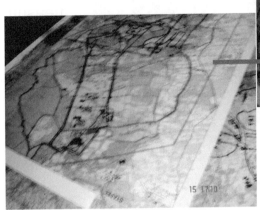

Check synchronisation via war-game to validate matrix

Figure 3.10   Coordination Measures captured on a whiteboard

Question 7. What Control Measures do I Need to Impose?

The fire control measures were developed by the BG Commander, to ensure that the indirect fire ordinance would not be placed on the advancing force elements. Five fire control lines were drawn on to an overlay on the paper map and numbered one to five. Each line was given a name, which was entered into the staff officer's notebook against the number used on the overlay (see Figure 3.11). The convention of naming phase lines was to ensure coordination between the force elements and indirect fire during the operational phase.

Summary of Observed Vignette and Comparison of Media Used

The record of the media used in this vignette is presented in Table 3.1, which indicates a variety of media including paper, maps, overlays, whiteboards, flipcharts and staff notebooks. Observation of the planning process suggests that the Combat Estimate method, media and products work well together. The plan was constructed within the 2 hour time frame, with only 17 minutes per question, and the staff officers had no difficulty using the conventional media. No difficulties were noted working between media, such as taking the Effects Schematic (question 3) from the flipchart and CoA (question 4) from the flipchart to produce the DSO (question 5) on an overlay. Similarly there were no problems noted for taking the DSO (question 5) from the overlay to produce the DSOM (question 6) on a flipchart. The point here is that translation between the media was straightforward, as all media and products were available for the staff officers to use at all times. The planning media and methods were not seen as a constraint on the planning process.

What control measures?

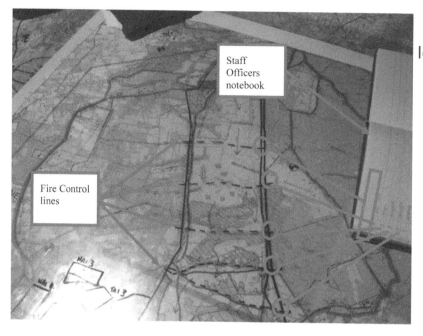

Staff Officers notebook

Fire Control lines

Identify control measures to help with co-ordination and help prevent fratricide

Figure 3.11   **Fire control lines on map and overlay also recorded in staff officer's notebook**

Table 3.1   **Media used during the planning process**

Media/Products	Paper	Maps/Overlays	Whiteboard	Flipchart	Staff Notebook
Warning Order	■				
Planning time line			■		
Q1. BAE/TI	■	■			
Q2. Mission Analysis	■		■		
Q2. CCIRs/RFI	■		■		
Q3. Effects Schematic	■			■	
Q4. COA	■			■	
Q4. DSO	■	■			
Q5. DSOM	■	■		■	
Q6. Wargame	■	■			
Q6. Co-ordination	■		■		
Q7. Fire control	■	■			■

The optimal choice of type and mode of communication is likely to be heavily dependent on the activity conducted. For some activities a textual document or a graphical image is more appropriate than a spoken alternative or vice-versa. The stage of any activity is also likely to heavily influence the optimal communication approach. Table 3.2 shows the degree of collaboration and cooperation required for different stages of the planning process. There is a clear divide; the latter stages of the process (questions four to seven) are best supported by collaboration (actors working individually with shared information). The earlier stages are much better suited to cooperative activity where the actors work together on one single product. Walker et al. (2009) report on the basic human communication structures seen inside a BG HQ, identifying eight key functions, some of which are comprised of further sub-functions. The eight key functions include the Higher Command Formation, the CO, CoS (CoS/2IC), the 'Principal' Planning Staff such as the IO/G2 (to varying extents it also requires the participation of individual roles such as Recce/ISTAR, Eng, A2/Log and Arty/AD. These have been placed on the periphery of the planning/principal staff node for illustration. There are also other ancillary command staff (such as those responsible for more general tasks and information management), that are referred to as sub-units (who are typically carrying out activities live in the battlespace) and, finally, the collection of graphics and planning aids derived from the Combat Estimate (artefacts that represent and transform information in some manner).

Walker et al. (2009) describe the human network as dynamic with different functional nodes and links becoming active under different activity stereotypes. The activity stereotypes that they identified were: providing direction (that is, the Commanding Officer (CO) directing communications and information outwards to subordinate staff in a prescribed and tightly coupled manner); reviewing (that is, the planning/principal staff communicate in a more collaborative manner with mutual exchange of information and ad-hoc usage of planning materials and outputs); and semi-autonomous working (that is, the HQ staff are working individually on assigned tasks

Table 3.2   Team work required for each stage of the planning process

Combat Estimate question	Task work or team work
Q1. What is the enemy doing and why?	Cooperative activity around the table
Q2. What have I been told to do and why?	Isolated intellectual activity followed by collaborative activity around the table
Q3. What effects do I want to have on the enemy?	Isolated intellectual activity followed by cooperative activity around the table
Q4. Where can I best accomplish each action/effect?	Collaborative activity in which the products are shared
Q5. What resources do I need to accomplish each action/effect?	
Q6. Where and when do the actions take place in relation to each other?	
Q7. What control measures do I need to impose?	

Social organisation

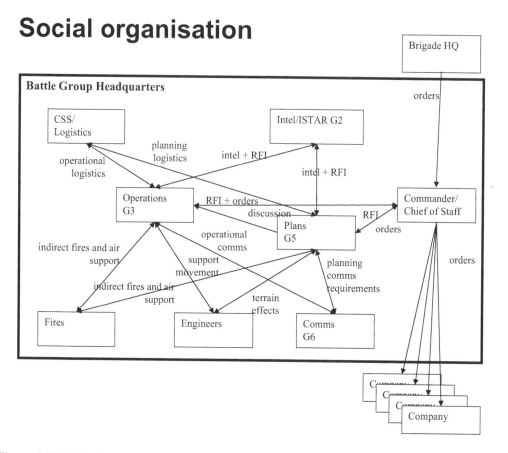

Figure 3.12   Relationships between the cells in Battle Group Headquarters during mission planning

and become relatively loosely coupled in terms of communication. The communication channels remain open but used in an ad-hoc, un-prescribed manner).

These basic structures account for most of the formal communications. The human network structure is complex, but some of the links are identified in Figure 3.12 above.

As Table 3.2 and Figure 3.12 indicate, the process of mission planning is a collaborative and cooperative process, both in terms of the contribution to the products and the verbal interactions. It is also very obvious that the planning team surrounds themselves with the planning artefacts. Maps, overlays, whiteboards and flipcharts literally adorn every surface. The plan is literally constructed in the space between these artefacts, as information is collected, transformed and integrated from the cognitive artefacts and the interactions between the planning team. The training that planners undergo reinforces the fact that the information needs to be 'public', for all to see and interact with. The planning process appears to focus on identifying the constraints (such as the mission, the enemy, the environment, the resources and assets) to help define the possible CoA. The process also requires an understanding of enemy doctrine and tactics to anticipate their likely behaviour and responses as well as military experience to know what effects are likely to achieve the desired outcome. Although it is difficult to quantify, there is certainly the opportunity for creativity in the way in which the plan is constructed. The planning team are continually trying to identify ways in which they can get the most from their finite resources and assets as well as preventing the enemy from anticipating their strategy. The planning process is also required to be flexible, as the planning process is continuous – as the process of issuing Fragmentory Orders (FRAGOs) suggests. Analysing the effects of digitising Mission Planning and Battlespace Management (MP/BM) is the subject of the rest of this book, with reference back to the analogue version of the same.

Chapter 4
Constraint Analysis

Complex socio-technical systems are notoriously difficult to evaluate, the relationships between measurable processes and high-level descriptions of performance are neither clear nor well understood. Consequently, it is extremely difficult to relate the changes made to physical components (such as the introduction of new technology) to overall system performance. The selection of performance measures for the evaluation of complex socio-technical domains is a challenging process. A wide range of conflicting factors, such as safety and expeditious flow, frequently influences these domains. These factors each have the potential to affect the overall performance of the system, commonly in unexpected ways. Further, even if appropriate measures are selected, and the correct data collected, challenges also exist in the synthesis of this information and the determination of its relative impact on overall system performance.

Using a systems-based modelling approach, this chapter will show how a multi-level system description can be used to describe system performance, and how these levels can be linked to relate low-level descriptions of components affordance to high-level descriptions of functional purpose.

The measures used in this analysis are derived from the Abstraction Hierarchy (AH; Rasmussen, 1985); part of the Cognitive Work Analysis (CWA) framework. CWA is a technique used to model complex socio-technical systems. The use of the framework to model military command and control is not a new concept, it has been successfully applied in the past (for example, Burns et al., 2004; Chin et al., 1999; Cummings & Guerlain, 2003; Jenkins et al. 2008a; Jenkins et al. 2008b; Jenkins et al., 2009; Lamoureux et al., 2006; Lintern et al., 2004; Naikar & Saunders, 2003). The framework is used to model different types of constraints, when used in its entirety it builds a model of how work can proceed within a given system. According to Vicente (1999), there are five separate phases, each phase considering different constraint sets. The analysis used in this chapter focuses on the initial phase, Work Domain Analysis (WDA). WDA provides a description of the domain or environment in which activity can take place. By focusing on constraints, an emphasis is placed on what is possible rather than what does, or should happen. The model is independent of specific actors and of activity. According to Crone et al. (2007), the CWA framework is important for several reasons. Firstly, it provides a way to categorise measures into meaningful groups. Secondly, it provides a mechanism to explore relationships between system components. The main tools used within WDA are the AH and Abstraction-Decomposition Space (Jenkins et al., 2009). The AH captures the purposes and the affordances of the examined system at a number of levels of abstraction (typically five). At the highest level, the overall functional purpose is considered. At the lowest level, the individual components that make up the system are examined. Using a 'why-what-how' relationship, these levels are linked together. Any node in the AH can be taken to answer the question of 'what' it does. The node is then linked to all of the nodes in the level directly above to answer the question 'why' it is needed. It is then linked to all of the nodes in the level directly below that answer the question 'how' this can be achieved. Through a series of means-ends-links, it is possible to model how individual components can have an influencing effect on the overall system purpose. Generally, although not exclusively, the upper four levels of the AH are independent of particular technology, they concentrate on the objectives of the work system and the means that can influence the objectives. It is primarily at the bottom row of the hierarchy where the individual components and technologies required are considered. Therefore, in most cases, where the computerised system conducts the same functions, the uppermost four levels of the AH are applicable to both new computerised systems and their legacy benchmarks.

The AH is a description of the technology that sits in the work system and the relations of that to superordinate purposes. At the base of the hierarchy, the physical functions of the system components can be examined. These functions can then be evaluated in terms of their contribution to the purpose-related functions. In turn, these can be compared to investigate their effect on the values and priority measures and finally on the overall functional purpose. By considering each level of abstraction in turn, and evaluating the performance of each part of the system, it is possible to understand which of the physical functions have a significant impact on the overall functional purpose. In this case, Subject Matter Expert (SME) opinion is used to evaluate the impact of each node in the hierarchy; each of the nodes within the AH is used to ask the question: is this part of the system significantly better, about the same, or significantly worse than the benchmark?

Method

An AH was constructed for the command process (see Figure 4.1). The document was created by the analysts using a number of resources: (1) a guide relating to the planning process trained to staff officers (the Combat Estimate Process; CAST, 2007) provided a clear description of the required information and products of each stage of the planning cycle. (2) Operational understanding was provided by a training video (BDFL, 2001) of a BG in a 'quick attack'. This was supplemented with the combined experience of the authors in military and non-military command and control domains (Stanton et al., 2008b; Jenkins et al., 2009). The final model (see Figure 4.1) was generated with the assistance of a military expert who was provided with an explanation of the AH framework. The expert was asked to consider the appropriateness of each of the nodes in the hierarchy as well as the links between them. As previously stated, with the exception of the bottom level (physical objects) the model is technologically agnostic. Essentially, the new technology has been introduced to perform the same processes as the legacy system in a different way. Instead of representing unit positions received by radio on a large paper map with stickers, the information is represented on a workstation with a digital map using automated reporting. Likewise, data that was shared through voice-based radio transmissions or physically distributed on pieces of paper in the legacy system can now be transmitted over a network. In the interpretation of this model, it is important to remember that the model is independent of activity and of actor, the focus is on what the system can afford, rather than the procedure required for achieving the affordance. Thus, a system with a standard radio can have the same affordance (communicate information on friendly positions) as an automated position reporting system. The model is also non-specific to any level of command; it represents the generic command and control process and is, therefore, applicable for both Brigade (Bde) and Battle Group (BG) in the analogue or the digital process.

At the highest level of abstraction, the functional purposes are listed. These capture the overall reason that the system exists. These purposes are independent of time; they exist for as long as the system exists. In this case, the overall purposes have been listed as 'Battlefield management' and 'Planning to enact higher command's intent'. These are considered the sole reasons for the command centre's existence.

At the second level down, the value and priority measures, a number of measures are captured for evaluating how well the functional purposes are being performed. The links in Figure 4.1 indicate which of the measures relate to each functional purpose. In this case, the majority of measures apply to both functional purposes with the exception of 'Fidelity of plans'; 'Fidelity of

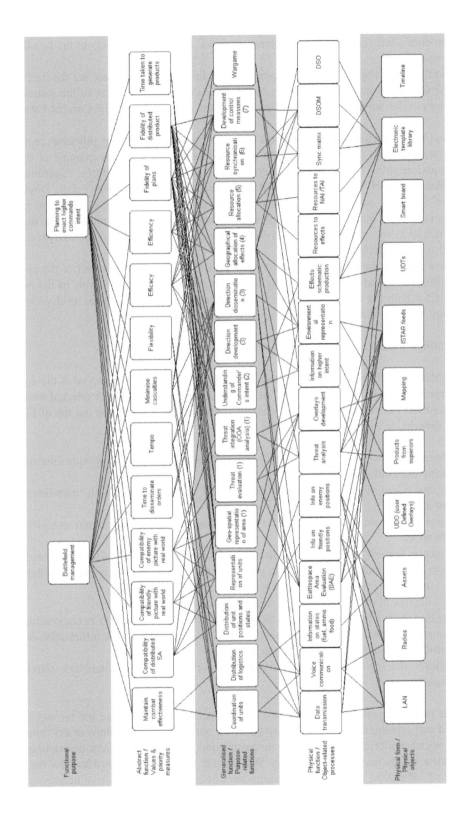

Figure 4.1   Abstraction Hierarchy for the command process

distributed product'; and 'Time taken to generate products', which relate solely to the functional purpose of planning.

In the middle of the diagram, the purpose-related functions are listed. These are the functions that can be performed, which have the ability to influence directly one or more of the values and priority measures. The numbers in brackets in some of the nodes relate to the Combat Estimate questions.

The second level from the bottom, the object-related processes, captures the processes required from the physical objects, in order to perform purpose-related functions. An example of this is 'Voice communication'; voice communication allows the system to perform 'Coordination of units'; 'Distribution of logistics'; 'Distribution of unit positions and states'; and 'Direction dissemination'. It should be noted that the model represents a model of what can happen, rather than what does or should happen; the emphasis with this approach is on constraints. At the object-related purposes level, the model concentrates on what the objects can physically do independently of the overall system functional purposes.

The bottom layer of the model lists the physical objects. This is the only part of the system that would differ for the analogue and the digital system. As the upper four levels of the model are applicable for both the analogue and the digital system it serves as an ideal mechanism for comparing how well the new digital system supports the command process.

Subjective Ratings of the Components Within the Abstraction Hierarchy

In total 11 SMEs (five from Battle Group (BG) and six from Bde) were surveyed at the end of the 3-week trial to establish their opinions of the new digital system. The SMEs selected were the senior officers from each of the cells of the command team (these are listed at the top of Figure 4.2). A short description of the roles is provided below.

- Intelligence officer – the role of the Intelligence officer is to predict and provide information on enemy disposition and activities.
- Operations (Ops) officer – the Ops officer controls and coordinates with 'own-forces' in current activities.
- Engineer – the role of the engineer is to provide information on terrain analysis and modification.
- Intelligence, Surveillance, Target Acquisition and Reconnaissance (ISTAR) – the role of this cell is to acquire, analyse and distribute products relating to intelligence and surveillance.
- Fires – the role of fires is to coordinate with artillery cells to direct fire.
- Plans – the role of the planning cell is to plan for future activities.
- Information-hub – this recently introduced role of the information-hub is responsible for coordinating and managing the technology within the headquarters. The information-hub is responsible for data management such as archiving.
- Logistics – the logistics cell is responsible for supporting the headquarters, providing equipment.

The participants were requested to take part in the experiment, the purpose and the use of the data provided was explained; participants were given the opportunity to withdraw at any time.

The AH model was briefly explained to the participants as a generic system description that was independent of any particular technology. Participants were asked to consider how the new technology had influenced their role and summarise whether the system is significantly better, significantly worse, about the same, or if the new system does not support the process.

	Legend	Battle Group					Brigade					
		Intelligence	OPS officer	Engineer	ISTAR	Fires	Intelligence	OPS officer	Engineer	Plans	Info-HUB	Logistics
	+ Significantly better **=** About the same **-** Significantly worse **N** Not supported											
FP	Battlefield management	+	+	+	+	+	-	+	+	+	+	+
	Planning to enact higher commands intent	-				-	-	=	-	=	=	=
Values and priority measures	Maintain combat effectiveness	N	+	N		+	=	=	+	-	=	+
	Compatibility of distributed situation awareness	+	+	+	+	+	=	+	+	+	-	=
	Compatibility of friendly picture with real world	+	+	+	N	+	-	+	+	+	-	+
	Compatibility of enemy picture with real world	+	-		N	=	-	=	+	+	-	N
	Time to disseminate orders	-	-	-	-	-	-	+	+	-		+
	Tempo	-	-	-	-	-	-	+			=	=
	Minimise casualties	-	+			+	-	+	+		=	=
	Flexibility	-	-		N	-	-	-	-			=
	Efficacy	-	-	-	-	-	=	+	=	-	=	+
	Efficiency	-	-	-	-	-	=	=		-		+
	Fidelity of plans	-	-	-	-	=	-	+	=	+		=
	Fidelity of distributed product	-	-	+	+	=	-	+	+	+		-
	Time taken to generate products	-	-		-	-	=					
Purpose related functions	Coordination of units		+		+	+	-	+	+	+	+	
	Distribution of logistics		+			+	=	+	+	=	=	=
	Distribution of unit positions and states	+	+	+	+	+	-	+	+	+	+	+
	Representation of units	+	-		+	+	-	+			+	+
	Geo-spatial representation of area (1)	+	+	+	-	-	+	-				
	Threat evaluation (1)		-		N	-	=			-	=	
	Threat integration (COA analysis) (1)		N		-		=	=	+	-	=	
	Understanding of Commander's intent (2)	=	=	=	N	=	=	=	=	+	+	=
	Direction development (3)		N			=	-	=	=	=	=	=
	Direction dissemination (3)	-	+	-	+	+	-	+	+	=	+	+
	Geographical allocation of effects (4)	N	N	N	-	N	=	=	-	-	=	+
	Resource Allocation (5)	N	N	N	N	-	=	=	+	-	=	=
	Resource Synchronisation (6)	N	N	N	N	=	+	+	+	+	=	+
	Development of control measures (7)	N	N	N	N	+	=	=	+	-	=	=
	Wargame	N	N	N	N	N	-		=	+	=	=
Object related processes	Data transmission	+	+	+	+	+	+	+	-	+	+	-
	Voice Communication	+	+	+	+	+	+	+	+	+	+	+
	Information on states (fuel, ammo food)		+		+	+	+	+	+		+	+
	Battlespace Area Evaluation (BAE)		-	-		-	-	=	-		=	
	Information on friendly positions	+	+	+	+	+	-	+	+		+	+
	Information on enemy positions	+			-		-	+	+		=	+
	Threat analysis	N	-		=	=	-	=	-		=	
	Overlays development	-	-	-		+	-	+	+	-	=	=
	Information on higher intent	-	=	N	=	-	=	+	+	=	=	=
	Environmental representation		-	=	-	+	+	-	-		-	=
	Effects schematic production		-		=	-	+	-			=	=
	Resources to effects	=		=	=		=			-	=	=
	Resources to Named/Target areas of interest	=		=	=		=	+		-	=	=
	Sync matrix		-		-	-	+	+	+	+	+	+
	Decision Support Overlay Matrix (DSOM)		+		+	=	-	+	+	=	+	
	Decision Support Overlay (DSO)		-			-	-			=		

Figure 4.2   Subjective opinion of the digital system in work domain terms (blank cells indicate that the respondent did not feel qualified to comment)

Participants were then read each of the lines in Figure 4.2 (representing each node from the upper four levels of the AH) from the bottom up (with the exception of the purpose-related functions section which was read in temporal order, from top to bottom) and asked to rate them.

Results

The results can be seen on the right-hand side of Figure 4.2. Each column represents the opinion of a single expert. A plus indicates that the participants rated the system as significantly better. A minus, that they rated it as insignificantly worse. An equals indicates they rated it as about the same. An 'N' indicates that they believed the new computerised system did not support this function. Finally a blank cell indicates that the respondent did not feel qualified to rate that part of the system.

The responses provided show that some aspects of the new system were considered to be significantly better, whilst others were considered significantly worse. This is most salient at the functional purposes level, where the overall opinion of the surveyed users was that the new system is detrimental to the planning process (significantly worse than the legacy system), whereas battlefield management was rated as significantly better. At the 'Values and priority measures' level, there are a number of nodes that scored as significantly worse than the legacy system. Notable examples include, time taken to generate products, tempo, efficacy and efficiency. The computerisation of the process seems to have slowed down the process of generating products, thus having a marked effect on planning. Conversely, aspects such as 'Compatibility of distributed situational awareness' and 'Compatibility of friendly picture with the world' receive ratings that indicate an improvement over the legacy system.

Using the means-ends-links identified in Figure 4.1 it is possible to explore further the relationships between ratings at different levels of abstraction. An analysis was conducted to investigate the level of concordance between ratings at different levels. Starting with the positive ratings, the responses of each respondent were taken in turn and a note made if there were concordant positive ratings at different levels as indicated by the means-ends-links. Figure 4.3 shows the results of this analysis. The links between nodes indicate the level of concordance, the link between 'Battlefield agreement' and 'Compatibility of distributed SA' indicates that eight out of a possible 11 respondents recorded both of these aspects of the system as significantly better, whereas the much thinner line between 'Battlefield management' and 'Efficiency' shows that only one respondent rated them both as positive. Thus, by following the thicker links a clear path can be made from improvements to affordances in 'Data transmission' and 'Voice communication', and overall battlefield management. Figure 4.4 shows the level of concordance for negative ratings; here strong links are made between planning and the traditional measures such as efficiency, flexibility and tempo. These can be related to the way individual units and the environment are represented, and in turn, linked to Decision Support Overlays (DSOs), Effects Schematic production and battlespace area evaluation. The results of this analysis suggest that the computerisation of this process offers little to assist the cognitive decisions relating to resource allocation, direction development and the allocation of effects. In fact, the digitisation of these processes is responsible for the negative responses in the values and priority measures.

In short, a clear link can be illustrated between the introduction of certain technology (encrypted radio and data transmission technologies) and the improvements in the functional purpose of battlefield management. Likewise, a clear link can be made between the introduction of technologies, such as the workstations, that change the environmental representation, overlay

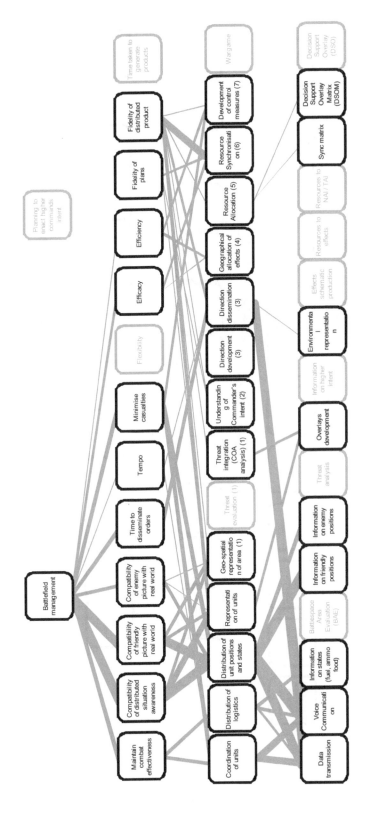

Figure 4.3 Concordance of positive ratings between levels (thickness of line indicates number of respondents making showing concordance)

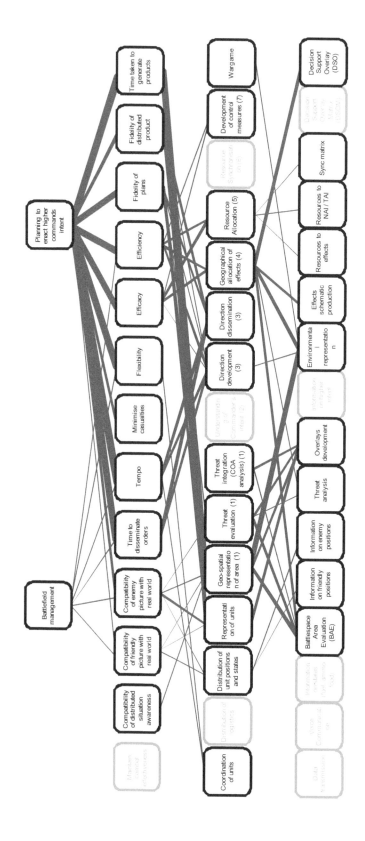

Figure 4.4   Concordance of negative ratings between levels (thickness of line indicates number of respondents making showing concordance)

development, Effects Schematic production and battlespace area evaluation, and the perception of degraded planning.

The evaluation of this system and the interpretation of the results indicates that the technology introduced to this system brings with it benefits, along with additional issues to be resolved. Elements of the new technology have clearly benefited battlefield management, showing improvements regardless of which level of abstraction the system is viewed from. Other aspects of the technology inserted have proved to have negative impacts on cognition and the time taken to generate planning products. These impacts are large enough to lead to a conclusion, from the surveyed experts, that the system's ability to support planning is significantly worse than the legacy system. This finding brings into question the suitability of the examined software tools designed to support planning, the results indicate a need to redesign these aids or reconsider their inclusion in the system.

Conclusions

The successful introduction of technology is a challenging issue, as Crandall, Klein & Hoffman (2006) point out:

> It is a mistake to overemphasise information management. The purpose of the information, and the IT, is to produce better and faster decisions and judgements, more effective planning, enhanced sensemaking, and so forth. Information management is a means it is not an end. If the information is well managed but does not have an impact on performance accomplishment, then the technology is without value – it's a toy, not a tool.
>
> (Crandall et al. 2006, p. 168)

In order to ensure the development of 'tools, not toys', it is essential that measures be developed for assessing the usefulness of a given technology. The changes to the way the physical affordances of a system are achieved need to be related to their effects on the systems functional purpose. This chapter introduces the concept of using the AH to generate such measures, and presents an approach for linking changes at a physical level to the overall functional purpose.

Whilst this method builds upon concepts from previous work, most notably the work of Naikar & Sanderson (2001) and Crone et al. (2003), the empirical approach has been applied in a far more expeditious manner. The hierarchical structure of this approach makes it appear similar to approaches based upon Hierarchical Task Analysis (HTA; Shepherd, 1985; Stanton, 2006) and Goal Directed Task Analysis (Goals, Operators, Methods and Selection rules (GOMS); Jones et al. 2003); however, unlike these approaches, the model produced by the WDA is not focused upon a specific actor conducting a specific task. The AH model is deliberately independent of actors, activities and situation. Thus, making it applicable for future systems configurations where work may be allocated differently between human and non-human elements. Further, the utility of the discussed approach lies in the relationships captured in the means-end-links.

Whilst the approach presented in this chapter has been to apply this technique for the retrospective evaluation of an existing system, it is contended that a similar approach could be adopted to predict and evaluate the potential impact of future system changes. The impact of changes to the way processes are conducted could be considered in terms of their effect on higher order functions. This approach is expected to be particularly beneficial in the early stages of reviewing potential system upgrade proposals the systematic approach could reveal, in broad terms, a prediction of an individual product's ability to influence a domain's function purpose(s). Further research is planned to explore the applicability of this extension of the described approach and how more objective measures of performance would enhance the approach.

Chapter 5
Hierarchical Task Analysis

Introduction to Hierarchical Task Analysis

Hierarchical Task Analysis (HTA; Annett et al., 1971) is a normative task analysis approach that is used to describe systems in terms of goals, sub-goals and operations. The 'task' in HTA is therefore something of a misnomer since HTA does not in fact analyse tasks, rather, it is concerned with goals (an objective or end state) and these are hierarchically decomposed. HTA works by decomposing activities into a hierarchy of goals, subordinate goals, operations and plans, which allows systems to be described exhaustively; it focuses on 'what an operator…is required to do, in terms of actions and/or cognitive processes to achieve a system goal' (Kirwan & Ainsworth, 1992, p. 1). HTA outputs therefore specify the overall goal of a particular system, the sub-goals to be undertaken to achieve this goal, the operations required to achieve each of the sub-goals specified and the plans, which specify the sequence, and under what conditions different sub-goals have to be achieved in order to satisfy the requirements of the superordinate goal.

Despite the vast range of Human Factors and Ergonomics methods available, the popularity of the HTA methodology is unparalleled. It is the most commonly used, not just out of task analysis methods, but also out of all Ergonomics methods (Annett, 2004; Kirwan & Ainsworth, 1992; Stanton, 2006). HTA has been applied now for over 40 years in all manner of domains and its heavy use shows no signs of abating, certainly not within Human Factors and Ergonomics circles. Although the process of constructing a HTA is enlightening in itself (that is, the analyst's understanding of the domain and system under analysis increases significantly), HTA's popularity is largely due to the flexibility and utility of its output. In addition to the goal-based description provided, HTA outputs are used to inform various additional Human Factors analyses; further, many Human Factors methods require an initial HTA as part of their data input (Stanton et al., 2005a). Indicative of its flexibility, Stanton (2006) describes a range of additional applications to which HTA has been put, including interface design and analysis, job design, training programme design and evaluation, human error prediction and analysis, team task analysis, allocation of functions analysis, workload assessment, system design and procedure design. The HTA process is rigorous, exhaustive and thorough, involving collecting data about the task or system under analysis (through techniques such as observation, questionnaires, interviews with Subject Matter Experts (SMEs), walkthroughs, user trials and documentation review to name but a few) and then using this data to decompose and describe the goals, sub-goals and tasks involved. The HTA procedure is presented in Figure 5.1.

The HTA analysis undertaken during the exercise involved the construction of HTAs for each component of the seven questions planning process. The HTAs were developed from the data collected during our observations of the activities undertaken during the exercise and also supplementary data taken from Standard Operating Instructions (SOIs) and interviews and discussions with SMEs.

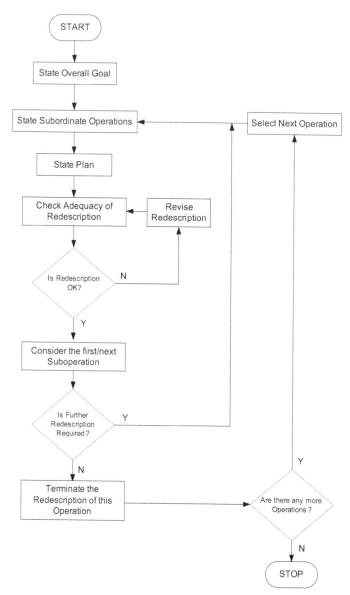

Figure 5.1   Hierarchical Task Analysis procedure

Digital Mission Planning and Battle Management Seven Questions Analysis

HTAs were constructed for the digital Mission Planning and Battlespace Management (MP/BM) component of each question during the seven questions planning process. The HTAs developed represent how each of the seven questions would be conducted using only the digital MP/BM system (as opposed to the combination of paper map process and digital MP/BM process that was observed during the exercise activities). Each HTA was then used to identify any design and usability issues relating to the digital MP/BM system and also for error identification purposes. The HTAs for each question are discussed in more detail overleaf.

Seven Questions Task Model

A high-level task model depicting the key activities undertaken was developed based on the seven questions digital MP/BM HTA. The task model also represents those seven questions products that can be constructed using the digital MP/BM system. The task model is presented in Figure 5.2.

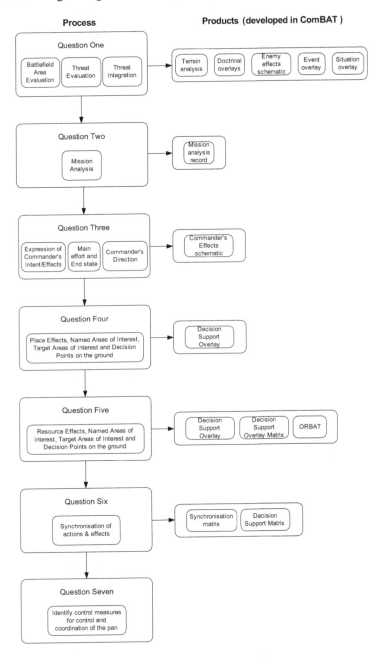

Figure 5.2   Combat Estimate Seven Questions task model

The task model shows, albeit at a high level, the critical activities that the Brigade (Bde) and Battle Group (BG) were engaged in during the seven questions planning process. Additionally, the task model shows the products that are produced at the conclusion of each question.

The digital MP/BM's role in the seven questions planning process is therefore to provide planners with the information that they require to undertake each component and also with the tools required to develop the planning products and distribute them accordingly. This analysis therefore focuses on the design and usability (that is, ease of use, potential for error, time required, user frustration and so on) of the different tools within digital MP/BM and also the provision of the information required for the planning process.

Question One

The process begins with question one, which involves the conduct of the BAE, the threat evaluation and the threat integration. The BAE involves conducting a terrain analysis using maps of the battlefield area and entails an assessment of the effects of the battlespace on enemy and friendly operations. It also involves the identification of likely mobility corridors, Avenues of Approach (AoAs) and manoeuvre areas. For the terrain analysis phase, the mnemonic 'OCOKA' is used, which comprises the following aspects of the terrain (CAST, 2007):

- **O**bservation
- **C**over and Concealment
- **O**bstacles
- **K**ey terrain and
- **A**oA.

Other key aspects of the terrain analysed during the terrain analysis component include the weather, restricted areas and potential choke points.

The threat evaluation phase involves identifying the enemy's likely modus operandi by analysing their tactical doctrine, past operations and their strengths and weaknesses. The end state of the threat evaluation phase is to 'visualise how the enemy normally executes operations and how the actions of the past shape what they are capable of in the current situation' (CAST, 2007, p. 12). Key aspects of the enemy that are investigated here include their strengths and weaknesses, their organisation and digital MP/BM effectiveness, equipment and doctrine and also their tactics and preparedness. The outputs of the threat evaluation phase are a series of doctrinal overlays which portray the enemy in terms of their organisation and digital MP/BM effectiveness, equipment and doctrine and also their tactics and preparedness.

The threat integration phase then involves combining the Battlefield Area Evaluation (BAE) and threat evaluation outputs in order to determine the enemy's intent and how they are likely to operate. The products of the threat integration include the enemy Effects Schematic, situation overlays for each enemy Course of Action (CoA) identified and an event overlay. Key elements identified during this phase include the Named Areas of Interest (NAIs) and likely enemy CoA. The output of the threat integration phase is the enemy Effects Schematic, which depicts the enemy's mission in terms of effects and intent, situation overlays, which depict likely and most dangerous enemy CoAs and an event overlay, which depicts when and where key tactical events are likely to occur.

The output derived from question one is therefore an understanding of the battlespace and its effects on how the enemy (and friendly forces) are likely to operate. The question one HTA is presented in Figure 5.3, Figure 5.4 and Figure 5.5.

Based on the HTA and also the evidence gathered during the exercise, we can conclude that undertaking question one with the digital MP/BM system is a difficult, error prone and overly time-consuming process. The process of analysing the battlefield area is made particularly difficult due to problems with the mapping, screen size and screen resolution on the digital MP/BM system. Firstly, the mapping and screen size capability during the exercise was such that the entire battlefield area could not be viewed in its entirety on the one screen. This meant that users had to continually zoom in and out of the battlefield area, a process which resulted in them losing context in terms of what area of the battlefield they were actually looking at.

Secondly, problems with the screen resolution and available mapping meant that users could not see specific areas (such as towns) in the level of detail required. It was difficult for users to establish an appropriate level of screen resolution on the system in order to be able to identify key areas (for example, towns, rivers, roads and so on) on the map. Thirdly and finally, producing overlays (for example, terrain analysis, event overlay) on the digital MP/BM system appeared to be time consuming and error prone. The drawing tools offered by the digital MP/BM system are convoluted and counter-intuitive (for example having to left click on the map and then right clicking on a drawing object whilst holding control in order to place objects on the map), and as a result the process of drawing on the map using the digital MP/BM system is difficult, time consuming and error prone.

As a result of these problems, planners used a combination of the old paper map process and the new digital MP/BM process. As certain aspects of the digital MP/BM system were too time consuming and deemed unusable, some of the activities were undertaken using paper maps, acetates and traditional drawing or marking-up tools (for example, pens, stickies and so on). Although this appeared to work well, there ultimately comes a point when the end planning product has to be produced on digital MP/BM so that it can be distributed around the planning system and this therefore leads to a doubling of the process (that is, the process is undertaken on paper and then undertaken again on the digital MP/BM system), which has the effect of lengthening the overall planning process and ultimately reducing operational tempo.

The question therefore remains as to whether the process of analysing the battlefield area (that is, question one) can be facilitated in any way via digitisation. Since it is a mainly a cognitive process it is questionable as to whether digitising this aspect of the planning process is beneficial in any way. The traditional paper maps certainly suffice in terms of providing an accurate representation of the battlefield area, and planners can 'zoom in and out' of the battlefield area using maps of different scale. Drawing on acetates is also much easier and user friendly than drawing on the digital MP/BM system. The outputs of question one consist of various overlays and textual descriptions and so it is here where perhaps the only real benefit of digitising the process emerges, since electronic outputs can be communicated more rapidly and to a wider audience. However, this could easily be achieved via the use of scanners or smartboards.

Question Two

Question two is known as the Mission Analysis and asks the question 'what have I been told to do and why?' Of specific interest during question two are the specified and implied tasks and the freedoms and constraints associated with the mission. Undertaking the Mission Analysis involves completing a Mission Analysis record, which requires a statement of the mission intent both 2 up (that is, 2 echelons up the command chain) and 1 up (1 echelon up the command chain), a statement of the main effort, specification of the specified and implied tasks, their deductions, Requests for Information (RFI) and Commander's Critical Information Requirements (CCIRs) and finally the freedoms and constraints associated with the mission. Specified tasks are typically found in the mission statement,

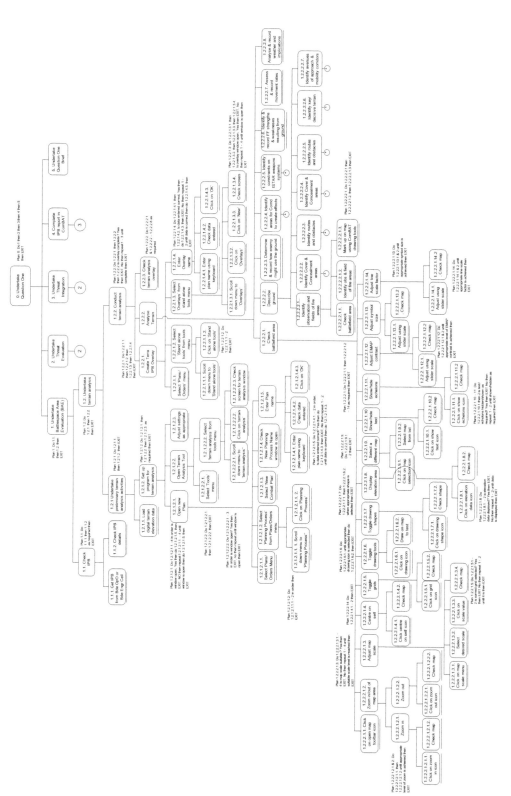

Figure 5.3   Question One HTA extract (1)

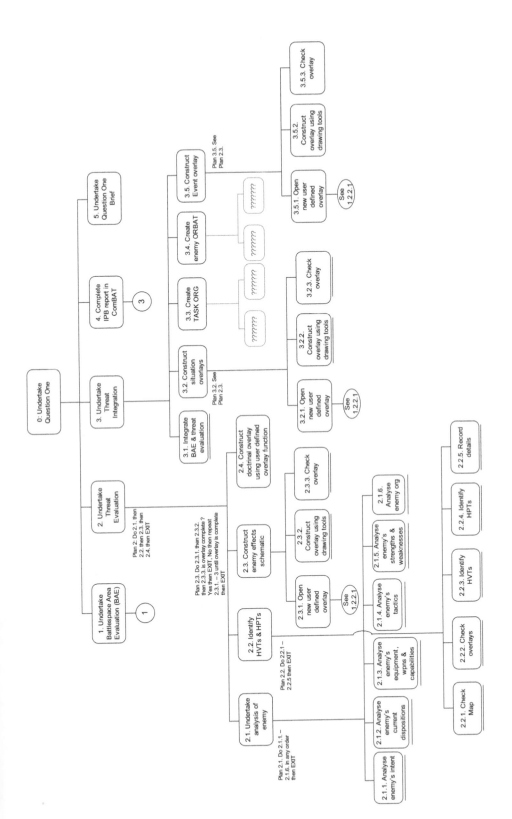

Figure 5.4   Question One HTA extract (2)

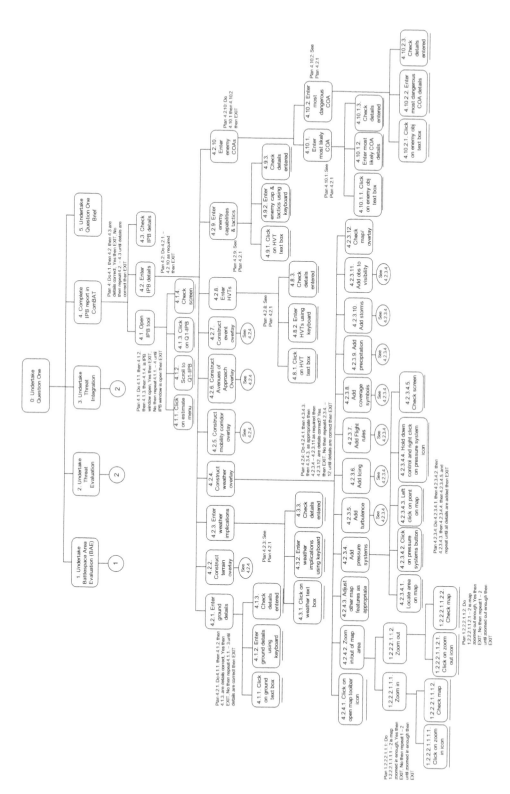

Figure 5.5   Question One HTA extract (3)

the coordination instructions, the Decision Support Overlay (DSO), the intelligence collection plan and the Combat Service Support for Operations (CSSO) (CAST, 2007). The output derived from question two is a completed Mission Analysis record detailing the mission, the main effort, the specified and implied tasks, any RFIs and the CCIRs.

The HTA for the digital MP/BM question two Mission Analysis process is presented in Figure 5.6. Completing the Mission Analysis component in digital MP/BM entails manually entering a description of the mission, the specified and implied tasks, any mission constraints and any additional information using the Mission Analysis tool. Following this, CCIRs and RFIs are entered into the RFI section in digital MP/BM. The Mission Analysis component of digital MP/BM uses a simple text box and keyboard entry system.

Completing the Mission Analysis record within digital MP/BM is a straightforward process and has the added benefit of allowing planners to review and refine the Mission Analysis details as they enter them. Again the key benefit here is that the Mission Analysis record can be quickly disseminated to other agents using the system's messaging functionality. The only problematic aspect revealed by the HTA is the lack of a free text entry function for the specified and implied tasks section. Currently the user has to use an 'add specified/implied task' button, which opens up a new window in which the details are subsequently entered. Again this is convoluted and allowing the user to enter the text directly into the specified/implied task text boxes would be more appropriate. Additionally (although this is a problem throughout the system), the use of a 'close' (X) icon, rather than an 'OK' icon to exit completed data entry windows is problematic and counter-intuitive.

Question Three

Question three involves the Commander specifying the effects that they wish to have on the enemy (CAST, 2007), what is referred to as their battle-winning idea, or 'that battlefield activity or technique which would most directly accomplish the mission' (CAST, 2007, p. 23). Based on the information derived from questions one and two, the Commander should now understand the battlespace area and the aims of the friendly forces involved and should also comprehend how the enemy are likely to operate. Using this understanding the Commander then identifies the effects required in order to achieve the mission and also prevent the enemy from achieving their mission. The Commander specifies their effects using an Effects Schematic and gives the purpose and their direction to the staff for each of the effects described. Additionally the Commander also specifies what the main effort is likely to be and also their desired end state. Additional direction designed to focus the planning effort is also given at this stage. This might include guidance on the use of applicable functions in digital MP/BM, principles of war and principles of the operation (CAST, 2007). Finally, the Commander confirms their CCIRs and RFIs.

During the exercise the completion of question three using the digital MP/BM system was not observed; a HTA was therefore developed based on the analyst's understanding of how question three should be undertaken using the digital MP/BM system. The HTA for question three is presented in Figure 5.7. The process involves the use of the user defined overlay and the digital MP/BM drawing tools to construct the commander's Effects Schematic. The tasks involved when opening a new user defined overlay and using the drawing tools to construct an overlay are described in the other HTAs in this section. We can therefore conclude that the same issues are present here, that is, that there are various usability problems associated with the drawing tools.

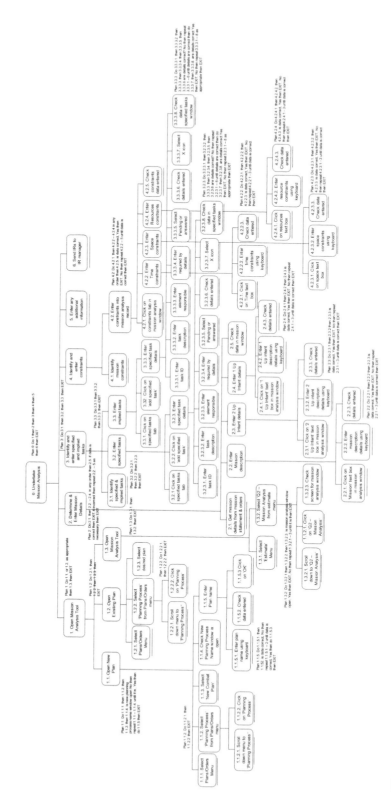

Figure 5.6   Question Two Mission Analysis HTA

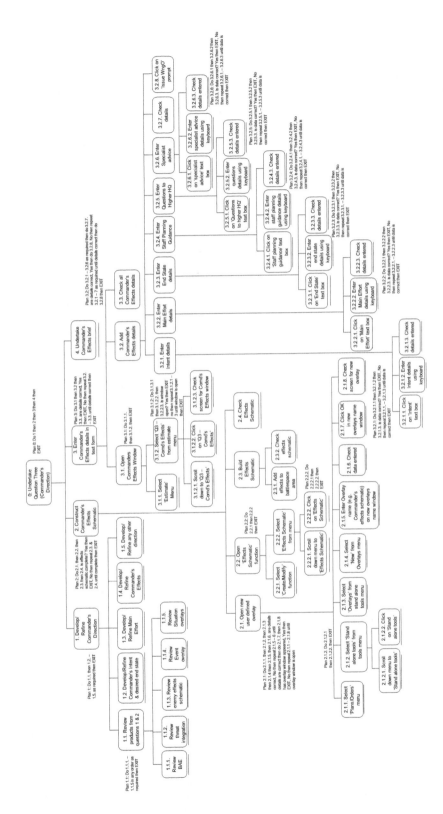

Figure 5.7   Question Three HTA

Questions Four, Five, Six and Seven

Questions four, five, six and seven are primarily concerned with the development of the CoAs required to achieve the Commander's desired end state. Since the questions are typically undertaken together or in parallel the digital MP/BM system contains a Q4–7 CoA development tool. The HTA for questions 4–7 is presented in Figure 5.8.

Question four involves identifying where each of the actions and effects specified by the Commander are likely to be best achieved in the present battlespace area and involves placing the Commander's effects, NAIs, Target Areas of Interest (TAIs) and Decision Points (DPs) on the map. Although some of the effects are likely to be dictated by the Commander and the ground, others, such as STRIKE and DEFEAT effects, can often potentially take place in a number of areas depending on a variety of factors such as enemy location, terrain and friendly force capability. The output of question four is the draft DSO which contains the Commander's effects, NAIs, TAIs and DPs for the mission. A HTA for the construction of the DSO is presented in Figure 5.9.

Undertaking question four in digital MP/BM involves manually adding the Commander's effects, NAIs, TAIs and DPs to the battlefield area and then textually adding these features using the DSO tool. Again due to the usability issues associated with the digital MP/BM system's drawing tools, producing the DSO in digital MP/BM is both time consuming and error prone. Observation of this process during one instance indicated that it took approximately four times longer (it took 60 minutes after significant practice) to produce the DSO using digital MP/BM as opposed to using paper maps and acetates.

Question Five

Question five involves specifying resources for each of the Commander's effects, NAIs, TAIs and DPs. This involves considering the effects required and then the mission, digital MP/BM power, type, size and strength of the enemy at each NAI and TAI. Much of this information can be derived from the assessment of the enemy's strengths and weaknesses made during question one as part of the threat evaluation. The output of question five is a series of potential CoAs for each effect, NAI and TAI and a Decision Support Overlay Matrix (DSOM). The Commander then makes a decision of how each effect, NAI and TAI is to be resourced, which leads to the production of the final DSOM. The DSOM is produced semi-automatically in digital MP/BM; however, some portions of it still require completion, namely the purpose, assets and remarks sections. The lack of explicit links between the NAIs, TAIs and DPs within the digital MP/BM DSOM is a problem and users cannot easily discern the relationship between the NAIs, TAIs and DPs. There is some concern over the automatic production of the DSOM and the lack of user involvement in this process may enhance error potential since the user is not refining the details as they complete the DSOM.

Question Six

Question six focuses on the time and location of each CoA, that is, when and where do the actions take place in relation to one another? To determine this, a synchronisation matrix is produced, which includes a statement of the overall mission and the concept of operations and then a breakdown of events related to time, including enemy actions, and friendly force components' activities and DPs. The output of question six is a draft synchronisation matrix and a decision support matrix.

Question six is completed within digital MP/BM using the synchronisation matrix (synch matrix) tool (manually the synch matrix is constructed using a flipchart). The digital MP/BM synch tool was found to be problematic during the exercise. For example, during the planning phase in

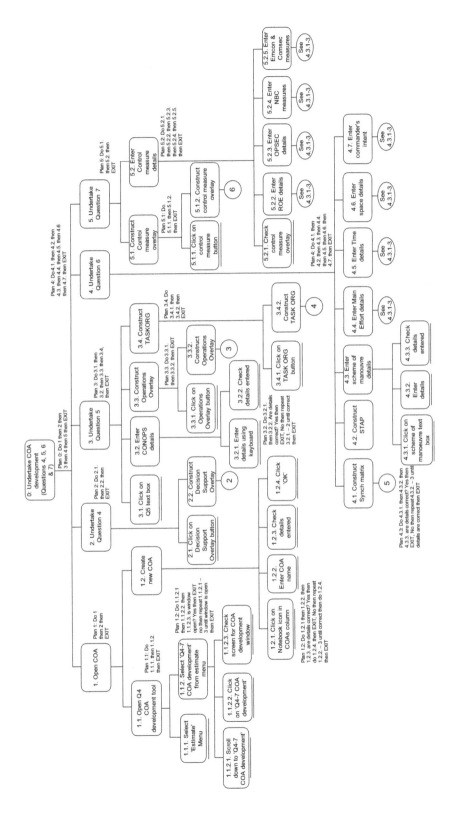

Figure 5.8 Questions Four–Seven HTA

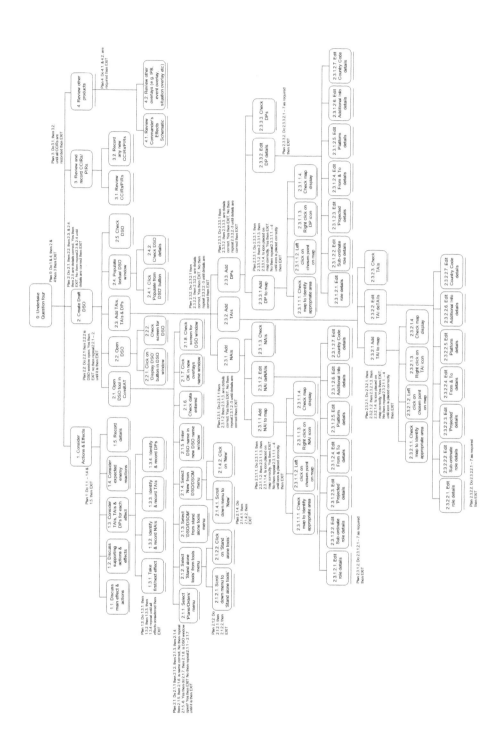

Figure 5.9 DSO construction HTA

BG the synch matrix was constructed on a digital MP/BM terminal in the plans cell. A HTA of the synch matrix construction task is presented in Figure 5.10.

Due to problems with the synch matrix tool, the synch matrix product took around 6 hours to complete. The process of constructing the synch matrix within digital MP/BM appeared to be unintuitive, error prone and overly time consuming. The old process involved manually drawing a synch matrix on a whiteboard and populating it as appropriate. The digital MP/BM process involves first constructing a synch matrix template and then populating it by adding different action groups (for example, enemy and friendly forces), actions (for example, recce, arty and so on) events and timings. The synch matrix is then distributed via publish and subscribe and printed out.

The user in this case made many errors and had to consult with various other users on how to undertake the process correctly. At one point (after approximately one hour) the user gave up and started over as they could not rectify an earlier error that they had made. The menu icons were also found to be unintuitive and the user had to continually float the mouse over the menu icons in order to see what they actually were. The errors made by the user range from failing to enter data (for example,

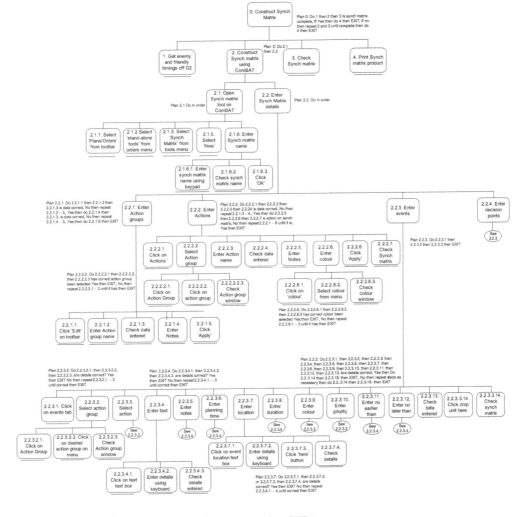

Figure 5.10　Synchronisation matrix construction HTA

failing to enter action group name), failing to select data (for example, failing to select action group or action), entering the same data twice (for example, entering action name twice), entering the wrong data (for example, selecting wrong item from drop down menu, entering wrong name) and constructing the overlay incorrectly (the user put all of the actions under the enemy rather than friendly action group and could not rectify it). Eventually the user had to stop and restart the entire process almost an hour after beginning it. This was due to the fact that the error in this case could not be recovered (although it would be simple to make it recoverable).

Also evident during this vignette was a lack of awareness as to what the digital MP/BM system is actually capable of. At one point a colleague pointed out that the synch matrix can be automatically populated (a process that was causing the user great difficulty) via the TASKORG product. Another colleague agreed that this could be done whereas the user constructing the synch matrix was not actually sure. However, all three did not know how to populate the synch matrix with the TASKORG. Printing the synch matrix product was also problematic as the printer used was incapable of printing the entire synch matrix on one page, which meant that the entire synch matrix could not be viewed at once. As a result of the problems identified above, the synch matrix was not ready and the orders were sent without the synch matrix.

It is apparent that the synch matrix construction process could be made much more simple and intuitive. For example, populating the Action group and Actions column of the table could involve simply clicking on the column and adding the action group or action name. Currently the user can right click on this column to modify or move existing action groups and actions but cannot add actions in this manner.

Question Seven

Finally, question seven involves identifying any control measures that are required for the CoAs specified. Control measures are the means by which activities are coordinated and controlled. Control measures include phase lines, boundaries, fire support coordination measures and lines, assembly areas and rules of engagement. Within the digital MP/BM system control measures are added to the map using the drawing tools and the details are entered textually within the Q4–7 CoA development window.

Human Error Analysis

The HTA outputs were used to inform the conduct of a Human Error Identification (HEI) analysis of the digital MP/BM software tool. HEI techniques offer a pro-active strategy for investigating human error in complex socio-technical systems. HEI works on the premise that an understanding of an employee's work task and the characteristics of the technology being used allow us to indicate potential errors that may arise from the resulting interaction (Baber and Stanton, 1996). Since a number of high-profile, human error-related catastrophes occurring in the late 1970s and early 1980s, such as the Three Mile Island, Bhopal and Chernobyl disasters, the use of HEI techniques has become widespread, with applications in a wide range of domains. HEI analyses typically offer descriptions of potential errors, their causal factors and consequences and proposed remedial measures designed to reduce the potential of the identified errors occurring.

Systematic Human Error Reduction and Prediction Approach

Of the many HEI approaches available, the Systematic Human Error Reduction and Prediction Approach (SHERPA) (Embrey, 1986), is the most commonly used and most successful HEI approach. Most of the approaches available are domain specific (that is, developed for a specific application within a specific domain); however, the SHERPA approach uses a generic error taxonomy and so can be easily applied in new domains. SHERPA was developed originally for use in the nuclear reprocessing industry but has since been applied in various domains.

SHERPA uses a generic External Error Mode (EEM) taxonomy linked to a behavioural taxonomy and is applied to a HTA of the task or scenario under analysis. The behavioural and EEM taxonomies are used to identify credible errors that are likely to occur during each step in the HTA. For example, each bottom level task step from the HTA is firstly classified as one of the five following behaviour types from the SHERPA behaviour taxonomy:

- Action – for example, pressing a button, typing in data and so on.
- Check – for example, making a procedural check.
- Information Retrieval – for example, retrieving information from a display or document.
- Information Communication – for example, talking to a colleague or co-worker.
- Selection – for example, selecting one alternative over another.

Each SHERPA behaviour classification has a set of associated EEMs. The SHERPA EEM taxonomy is presented in Figure 5.11.

The EEM taxonomy and domain expertise are then used to identify, based on the analyst's subjective judgement, any credible error modes for the task step in question. For each credible error identified, a description of the form that the error would take is provided, such as 'pilot dials in wrong airspeed' or 'pilot fails to check current flap setting'. Next, the analyst describes any consequences associated with the error and any error recovery steps that would need to be taken in event of the error. Ratings of ordinal probability (Low, Medium or High) and criticality (Low, Medium or High) are then provided. The final step involves specifying any potential design remedies (that is, how the interface or device can be modified in order to remove or reduce the chances of the error occurring) for each of the errors identified. A flowchart depicting the SHERPA procedure is presented in Figure 5.12.

A SHERPA analysis was conducted using the HTAs presented previously. This involved one analyst predicting errors for each of the bottom level task steps contained in the seven questions planning HTAs. For example purposes, an extract of the SHERPA analysis for the question one HTA is presented in Table 5.1 .

Conclusions

HTA Analysis

The HTA analysis indicated that there are significant problems associated with the seven questions-related planning tools that the digital MP/BM system offers. In the main, the majority of these tools are counter-intuitive, difficult and time consuming to use and error prone. Furthermore, when compared to their paper map process counterparts, it appears that the digitised versions offer no real benefit (the only significant benefit being the ability to disseminate planning products quicker, further and to a wider audience; however this is tempered by the additional time taken to complete

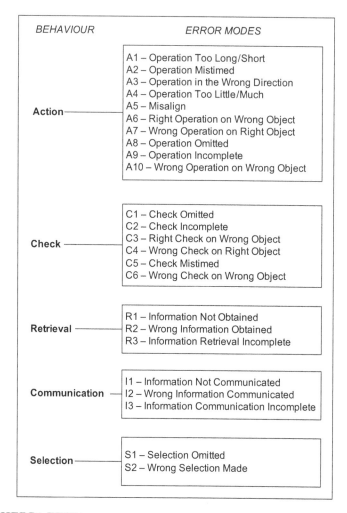

Figure 5.11   SHERPA EEM taxonomy

the planning products using the digital system). A summary of the main findings in relation to the planning products and tools used during the seven questions process is presented in Table 5.2.

The HTA analyses therefore indicate that there are significant problems associated with the user interface and tools within the digital MP/BM system. The more pertinent issues are summarised below:

- *Lack of standardised conventions*. The most striking finding to emerge from the HTA analysis is the general lack of standardised user conventions contained within the digital MP/BM system. The designers of the GUI have failed to exploit standardised user conventions from existing (and well used) systems, such as Windows XP, Microsoft Visio (drawing package) and Microsoft Word. Examples of these omissions include the so-called digital MP/BM copy process (which involves holding control on the keyboard and right clicking the mouse to copy, and then left clicking to paste), having to click on X in some windows upon completion of a task and the lack of a drag and drop drawing function. These processes all

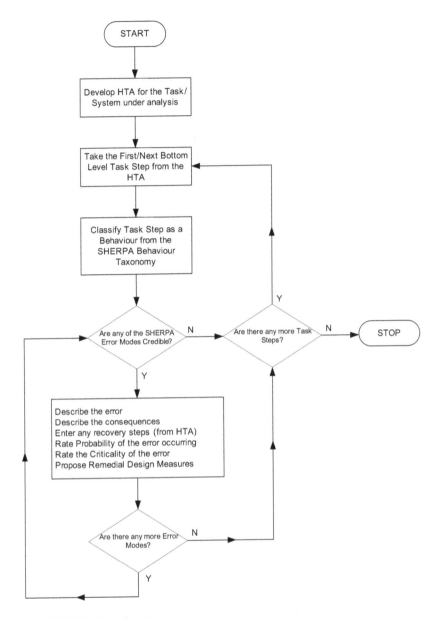

Figure 5.12 SHERPA flowchart

represent instances of where standardised conventions have been overlooked and are thus often alien to new users. Also the icons used (that is, for buttons on toolbars) are typically not standard ones and are often confusing.

- *Convoluted processes*. A number of the processes involved when using the digital MP/BM tool are convoluted. In particular, the marking-up of maps and the development of overlays are two processes that are overly complicated and require intricate task steps. Both represent instances where the designers have failed to exploit the user's mental model of the current process and also standardised conventions.

Table 5.1 Question SHERPA analysis extract

Task	Error Mode	Error Description	Consequence	Recovery	P	C	Remedial Measures
1.2.1.1. Load digital terrain elevation data	A8 – Operation Omitted	User forgets to load digital terrain elevation data	Digital terrain elevation data is not loaded	1.2.2	M	M	- Digital terrain elevation prompt
1.2.1.2.1.1.1. Scroll down menu to 'Planning process'	A4 – Operation too much/too little	User scrolls too far down the Plans/Orders menu	Wrong menu function is highlighted/selected	Immediate	M	L	- Customised menus
1.2.1.2.1.1.1. Scroll down menu to 'Planning process'	A4 – Operation too much/too little	User does not scroll far enough down the Plans/Orders menu	Wrong menu function is highlighted/selected	Immediate	M	L	- Customised menus
1.2.1.2.1.1.1. Scroll down menu to 'Planning process'	A5 - Misalign	User scrolls down the Plans/Orders to the wrong menu item	Wrong menu function is highlighted/selected	Immediate	M	L	- Customised menus
1.2.1.2.1.3. Select 'New Combat Plan'	A5 - Misalign	User moves mouse off the menu	Menu is no longer displayed	Immediate	M	L	- Mouse click to remove menus rather than mouse over
1.2.1.2.1.3. Select 'New Combat Plan'	A6 – Right operation on wrong object	User selects the wrong item from the Planning process menu	The wrong plan is opened by the system	Immediate	M	L	N/A
1.2.1.2.1.4.1. Enter plan name using keyboard	A6 – Right operation on wrong object	User mis-spells or enters the wrong plan name in the new plan name window	Wrong plan name is entered	Immediate	M	L	N/A
1.2.1.2.1.4.2. Check data entered	C1 – Check omitted	User fails to check plan name entered	Wrong plan name may be entered	Immediate	M	L	- Prompt to check plan name details
1.2.1.2.1.4.3. Click on OK	A5 - Misalign	User clicks on cancel	Plan name data is lost	Immediate	L	H	- Larger, clearer OK and Cancel icons
1.2.1.2.1.4.3. Click on OK	A6 – Right operation on wrong object	User clicks on X icon in the top right hand corner of the screen	ComBAT system begins shut down process	Immediate	H	H	- Increased consistency in the interface (X icon should not be used as an OK button!)

Table 5.2   HTA analysis product construction findings summary

Planning Product	Conclusion	Remedial Measures
BAE	- Process of marking up map convoluted - Overly time consuming and difficult - Poor map legibility - Screen size limits context - Unintuitive drawing tools	- Drag and drop drawing function - Improved screen resolution - Improved mapping - Map within map display
Mission Analysis	+ Straightforward process + Completion of table ensures data check - Lack of a free text entry function for specified/implied tasks	- Free text entry function for specified and implied tasks
Effects schematic	- Drawing difficult, time consuming and convoluted - Drawing tools unintuitive - Poor map legibility/screen resolution - Screen size limits context	- Drag and drop drawing function - Improved screen resolution - Improved mapping - Map within map display
Overlays	- Process of marking up map convoluted - Overly time consuming and difficult - Poor map legibility - Screen size limits context - Unintuitive drawing tools	- Drag and drop drawing function - Improved screen resolution - Improved mapping - Map within map display
DSO	- Time consuming and difficult	- Drag and drop drawing function - Improved screen resolution - Improved mapping - Map within map display
DSOM	+ Auto-construction	
Synch Matrix	- Time consuming and difficult - Display limits viewing	- Free text synch matrix construction - Logical presentation
Task ORG	- Poor layout (user cannot see entire Task ORG on one screen) - Overly time consuming and difficult - Convoluted process - Error prone	- Use columns like traditional process - Drag and drop function - Logical presentation

- *Oversized menus.* Some of the menus contained within the digital MP/BM system are overly large and contain too many items to be usable. This increases user interaction time and also the potential for error. For example, the create/modify menu contains 38 different menu items.
- *Drawing tools unintuitive and overly complex.* The drawing system offered by the digital MP/BM system is especially problematic. The system is unintuitive, difficult to use and heavily prone to error. As a consequence, drawing processes (for example, marking-up of maps, overlay construction) is time consuming and complex; so much so that the paper map drawing system was the preferred option during the exercise activities observed.

- *Lack of consistency*. In terms of user interface design, consistency refers to 'common action sequences, terms, units, layouts, colour, typography and so on within an application program' (Schneiderman, 1998, p. 13). Schneiderman (1998) suggests that consistency is a strong determinant of the success of systems and includes it in his eight golden rules of interface design. At times there appears to be a lack of consistency between different aspects of the digital MP/BM system's interfaces. For example, in some cases the user can click on 'OK' buttons to finish a process, where in other instances there are no 'OK' buttons and the user has to click on 'X' to signal that they have completed a process.
- *Inappropriate or lack of use of automation*. Currently the system does not appropriately exploit the capability for automating certain aspects of the planning process. Processes such as the loading up of the system and the Mission Analysis could easily be improved through the use of appropriately designed automation.
- *Interface clarity*. The current GUI is insufficient to support quick, intuitive and error free performance. Certain aspects of the interface are not sufficiently clear or prominent and as a result, it may take new users a considerable amount of time to find them or to determine what they actually are.

To summarise the main findings derived from the HTA assessment in relation to the overall digital MP/BM system, the advantages and disadvantages of both the paper map and the digital MP/BM planning process can be compared at a high level. This is represented in Figure 5.13.

Systematic Human Error Reduction and Prediction Approach Analysis

The SHERPA analysis highlighted a number of different errors that are likely to occur as a result of user-digital MP/BM interactions. A summary of the different error types is given below:

- *A4 Operation too little/too many errors* – mainly involved not scrolling enough or scrolling too far down the digital MP/BM menus. This resulted in the user selecting the wrong item from the menu.
- *A5 Misalign errors* – mainly involved the user selecting the wrong area on the map display when marking-up maps, selecting the wrong item from menus or pressing the wrong button or command on the interface.
- *A6 Right operation on wrong object errors* – there were many A6 errors identified. These included the user selecting the wrong function from the toolbar or drop down menus, clicking on the wrong item/button/command on the interface (due to inadequate design of icons), entering the wrong data or pressing the wrong key.
- *A7 Wrong operation on right object errors* – these errors mainly included incorrect data entry errors.
- *A8 Operation omitted errors* – these errors involved the user failing to perform a task such as failing to save the current file, failing to enter data, failing to select required functions and failing to locate tools and functions.
- *C1 Check omitted errors* – these errors involved the user failing to check data of some sort, including failing to check data entered, failing to check data presented by digital MP/BM and failing to check that data had been copied successfully.
- *R2 Wrong information obtained errors* – these errors were mainly misread errors where the user misreads data that is presented by the system.

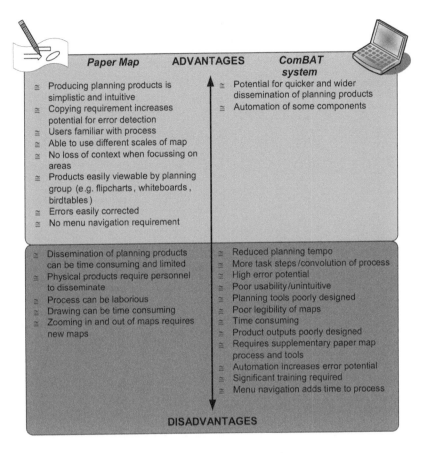

Figure 5.13 Advantages and disadvantages of each planning process

- *S2 Wrong selection errors* – involved the user making a wrong selection of some sort, such as selecting the wrong function from the tool bar or drop down menus and the user selecting inappropriate positions.

A series of remedial measures were proposed in order to eradicate or reduce the potential of the identified errors occurring.

Chapter 6
Distributed Situation Awareness

Digital MP/BM and Distributed Situation Awareness

The concept of Situation Awareness (SA) focuses on how operators in complex systems acquire and maintain sufficient awareness of 'what is going on' (Endsley, 1995a) in order to achieve safe and efficient task performance. Its importance is such that it is now widely recognised to be a critical consideration in system design and evaluation (for example, Endsley,1995a; Endsley, Bolte & Jones, 2003; Salmon et al., 2009; Shu & Furuta, 2005).

Unsurprisingly, the concept of SA emerged within a military context (Endsley, 1995a), where it is an integral component that can make the difference between life and death and victory and defeat in conflicts. SA is a critical commodity in the military land warfare domain, where distributed teams have to understand dynamic, information rich, uncertain, rapidly changing environments and elements and plan and execute activities against multiple adversaries working to defeat them. For example, Rasker et al. (2000; cited in Riley et al., 2006) point out that command and control teams need to perceive, interpret and exchange large amounts of ambiguous information in order to develop and maintain the SA required for efficient decision-making performance. Further, unlike civilian domains, military systems also have the added complexity of an adversary attempting to inhibit the development and acquisition of SA during operations. During mission planning, inadequate or erroneous SA can ultimately lead to inadequate or inappropriate plans and Courses of Action (CoA) and during battle execution degraded SA can lead to loss of life, failed missions, and in the worst case, loss of overall conflicts. In a military context, SA has been defined as:

> 'the ability to have accurate real-time information of friendly, enemy, neutral, and non-combatant locations; a common, relevant picture of the battlefield scaled to specific levels of interest and special needs.'

> (TRADOC, 1994; cited in Endsley et al., 2000).

In relation to digital mission support tools, Endsley & Jones (1997) suggest that the way in which information is presented influences SA by determining how much information can be acquired, how accurately it can be acquired, and to what degree it is compatible with SA needs. Prior to the field exercise that is the focus of this book, one of the critical success factors cited by Macey (2007b) was that the digital MP/BM system enhances SA during mission planning and execution activities. In addition, a number of claims have been made regarding the digital MP/BM system with regard to the impact on SA that it will have during operations. For example, in relation to SA the systems creator suggests that the system will facilitate:

- greater timeliness of the passage of information;
- greater accuracy in the passage of information;
- improved SA of own position;
- improved SA of friendly positions;

- improved SA of enemy positions; and
- improvements in the speed and accuracy with which information can be sent, processed and acted upon.

Therefore how SA-related information is assimilated, what information is presented by the system and to whom, the format in which is it presented and the accuracy and timeliness of its presentation to users are pertinent issues when considering the assessment of the digital MP/BM system. All of these issues ultimately relate to the overall level of SA that is afforded by the digital MP/BM system. This chapter presents an evaluation of the digital MP/BM system's design from the perspective of user, team and system SA.

Distributed Situation Awareness

Distributed Situation Awareness Model

Although various viewpoints on SA exist, ranging from individual SA 'in-the-head' models (for example, Endsley, 1995) to team 'shared SA' models (for example, Endsley & Robertson, 2000), it is our view that, within complex socio-technical systems such as military land warfare ones, it is pertinent to take the entire system, including the human operators and the technological agents that they use, as the unit of analysis and to focus on the so-called Distributed Situation Awareness of the system as a whole, rather than the SA of each individual unit working within the system. This is of course in line with wider movements within Human Factors toward systems level analysis (for example, Hollnagel, 1999; Hutchins, 1995). From this viewpoint it is the system that collectively holds the SA required for collaborative task performance.

Describing SA from a distributed cognition type perspective is nothing new; in the late 1990s Artman & Garbis (1998) first called for a systems perspective model on SA, suggesting that the predominant individualistic models of that time were inadequate for studying SA during teamwork. Instead they urged a focus on the joint cognitive system as a whole from a distributed cognition (Hutchins, 1995) viewpoint. Artman and Garbis (1998) subsequently defined team SA as 'the active construction of a model of a situation partly shared and partly distributed between two or more agents, from which one can anticipate important future states in the near future' (Artman & Garbis, 1998, p. 2). Following this, Stanton et al. (2006) laid the foundations for a theory of Distributed Situation Awareness in complex systems, and these foundations were built upon by Stanton et al. (2009) who outlined a Distributed Situation Awareness model, developed as a result of applied research in a range of military and civilian command and control environments. It is the Distributed Situation Awareness model outlined by Stanton and colleagues that forms the basis for the assessment presented in this chapter.

The Distributed Situation Awareness model is underpinned by four theoretical concepts: schema theory (for example, Bartlett, 1932), genotype and phenotype schema, Neisser's (1976) perceptual cycle model of cognition and, of course, Hutchin's (1995) distributed cognition approach. Following Hutchins (1995) and Artman & Garbis (1998), the model takes a systems perspective approach to SA and views SA as an emergent property of collaborative systems; SA therefore arises from the interactions between agents and is associated with agents but does not reside within them as it is borne out of the interactions between them. At a systemic level, awareness is distributed across the different human and technological agents involved in collaborative endeavour. Scaling the model down to individual team members, we suggest that team member SA represents the state of their perceptual cycle (Neisser, 1976); individuals possess genotype schema that are triggered by the task relevant nature of task performance, and during task performance, the phenotype schema comes to the fore. It is this task and schema driven content of team member SA that brings the shared SA notion into question. Rather than possess shared SA (which suggests that team members understand a situation or elements of a situation in the same manner), our model instead suggests that team members possess unique, but compatible, portions of awareness. We argue that team members experience a situation in different ways as defined by their own personal experience, goals, roles, tasks, training, skills, schema and so on. Compatible awareness is therefore the phenomenon that holds distributed systems together. Each team member has their own awareness, related to the goals that they are working towards. This is not the same as other team members, but is such that it enables them to work with adjacent team members. Although different team members may have access to the same information, differences in goals and roles, the tasks being performed and their schema mean that their resultant awareness of it is not shared; the situation is viewed differently based on these factors. Each team members' SA is however compatible since it is different in content but is compatible in that it is all collectively required for the system to perform collaborative tasks successfully. It is this compatibility that allows each component of the system to connect together.

The question remains as to how Distributed Situation Awareness is built between team members. Of course, the compatible SA view does not discount the sharing of information, nor does it discount the notion that different team members have access to the same information; this is where the concept of SA 'transactions' applies. Transactive SA describes the notion that agents within collaborative systems enhance the awareness of each other through SA 'transactions'. A transaction in this case represents an exchange of SA information from one agent to another (where agent refers to humans and artefacts). Team members may exchange information with one another (though requests, orders and situation reports); the exchange of information between team members leads to transactions in the SA being passed around; for example, the Request for Information (RFI) gives clues to what the other agent is working on. The act of reporting on the status of various elements tells the recipient what the sender is aware of. Both parties are using the information for their own ends, integrated into their own schemata, and reaching an individual interpretation. Thus the transaction is an exchange rather than a sharing of awareness. Each agent's SA (and so the overall Distributed Situation Awareness) is therefore updated via SA transactions. Transactive SA elements from one model of a situation can form an interacting part of another without any necessary requirement for parity of meaning or purpose; it is the systemic transformation of situational elements as they cross the system boundary from one team member to another that bestows upon team SA an emergent behaviour.

Modelling Distributed Situation Awareness

Viewing SA as a systems level phenomenon has interesting connotations for its assessment. Individual-based measures such as SAGAT (Endsley, 1995b) and SART (Taylor, 1990) are not applicable since they focus exclusively on the awareness 'in-the-head' of individual agents and overlook the interactions between them. Instead what is required is an approach that is able to describe the concept from a systems perspective; this includes the information that is distributed around the system, the usage of this information by different system elements, how the information is combined together to form 'awareness' and most importantly the SA-related interactions between system elements.

We put forward the propositional network methodology as a way of describing a systems SA, since it depicts, in a network, the information underlying a system's knowledge, the relationships between the different pieces of information and also how each component of the system is using each piece of information. The approach has been applied to a number of real world collaborative scenarios, including naval warfare (Stanton et al., 2006), railway maintenance operations (Walker et al., 2006), energy distribution substation maintenance scenarios (Salmon et al., 2008) and military aviation airborne early warning systems (Stewart et al., 2008). A propositional network is essentially a network depicting the information underlying a system's awareness and the relationships between the different pieces of information. They represent Distributed Situation Awareness as component information elements (or concepts) and the relationships between them, which relates to the assumption that knowledge comprises concepts and the relationships between them (Shadbolt & Burton, 1995). Depicting a system's awareness in this way permits the representation of the usage of different pieces of information by different agents (human and non-human) within the system and also the contribution to the systems awareness by different agents.

Propositional networks can be constructed from a variety of data sources, depending on whether Distributed Situation Awareness is being modelled (in terms of what it should comprise) or assessed (in terms of what it did comprise). These include observational or verbal transcript data, Critical Decision Method (CDM; Klein & Armstrong, 2004) data, HTA (Stanton, 2006) data or data derived from work-related artefacts such as Standard Operating Instructions (SOIs), user manuals, procedures and training manuals. The first step involves taking the input data and defining the concepts, or information elements, followed by the relationships between them. For this purpose, a simple content analysis is performed on the input data (for example, verbal transcript, CDM responses and so on) and keywords are extracted. These keywords represent the information elements, which are then linked based on their causal links during the activities in question (for example, contact 'has' heading, enemy 'knows' plan and so on). The output of this process is a network of linked information elements; the network contains all of the information that is used by the different agents and artefacts during task performance and thus represents the system's awareness; the network (information elements + relationships between them) represent what the system 'needed to know' in order to successfully undertake task performance. Information element usage is also defined and represented via shading of the different nodes within the network. Thus, the information elements related to the Distributed Situation Awareness and also the usage, ownership and sharing of these information elements as the scenario unfolds over time can be defined.

To demonstrate how propositional networks are constructed, Figure 6.1 presents an extract of a verbal transcript collected during a study of Distributed Situation Awareness during land warfare activities; the keywords extracted from this transcript via content analysis are highlighted in bold and the ensuing propositional network is presented on the right-hand side of the figure.

Distributed Situation Awareness Assessment

For the purposes of this analysis, we were specifically interested in the impact on Distributed Situation Awareness that the digital Mission Planning and Battlespace Management (MP/BM) system had and also how effectively the digital MP/BM system supported the SA requirements of its different users. Based on the literature (for example, Endsley & Jones, 1997; Riley et al., 2006), it was possible to postulate a number of critical SA-related requirements that electronic mission planning and execution systems should satisfy in order to enhance, rather than inhibit, SA acquisition and maintenance during planning and execution activities. These are summarised below:

1. *Support for the range of different user SA requirements.* Military planning and execution activities are undertaken by various agents with multiple (some common and some entirely

Propositional Network

Verbal Transcript

*"the goal for the **terrain** analysis part of **question one** is to produce an **overlay** to **brief** the **commander** on the military **effects** of the **terrain** on his **mission**. That includes **areas** that are **restricted** or **very restricted** or their **movement** is **restricted** or **very restricted**... it includes **elevation**, **height**....then it goes into detail regarding **observation** & **fields of fire**, **cover** and **concealment**, **obstacles**, **key terrain** and **avenues of approach** which is basically **movement corridors** that can be used both by the **blue forces** in achieving the **mission** but also approach **routes** by the **enemy** as well".*

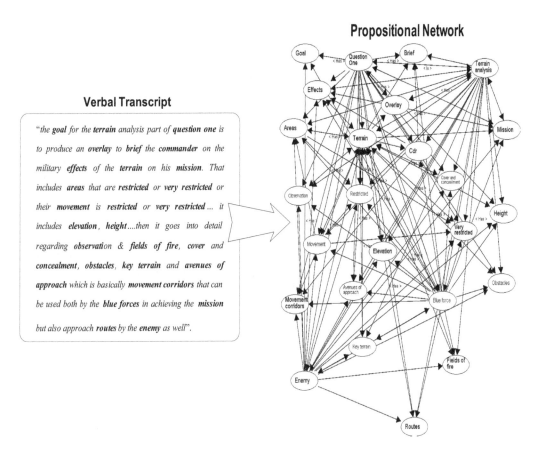

Figure 6.1   Propositional network example; figure shows verbal transcript and the resultant propositional network that is constructed based on the identification (via content analysis) of keywords from the verbal transcript data

different) goals. Digital mission support systems therefore need to be able to present the appropriate information, in an appropriate format, to the appropriate users at the appropriate time. (That is, does the system present the appropriate information required for planning and execution, in a format amenable to rapid understanding and also does it present it to the appropriate people?). Endsley & Jones (1997), suggest that the key to SA and information dominance in future warfare is in getting the right information to the right person at the right time in a form that they can quickly assimilate and use.

2. *Presentation of SA-related information in a timely manner*. Mission planning and execution activities are highly time critical and operational tempo is one of the key factors in the success of land warfare missions. Distributed Situation Awareness-related information should therefore be presented to users in a timely manner, without any delay, in order to enhance planning and execution tempo.

3. *Development and dissemination of planning products in a timely manner*. Again focussing on the impact on operational tempo, planning products need to be developed and disseminated in a timely manner (Riley et al., 2006). The system should therefore support the rapid development and distribution of plans and planning products.

4. *Presentation of accurate SA-related information.* An efficient level of Distributed Situation Awareness is dependent upon the exchange of accurate SA information between agents within the collaborative system. Likewise, efficient planning and battle execution is contingent upon an accurate understanding of the situation. It goes without saying that the information presented by the mission planning and execution system should be up-to-date and accurate. Bolia et al. (2007) point out that inaccurate data can emerge from erroneous assumptions made by data fusion algorithms (for example, a data fusion algorithm deciding that two sensor inputs represent a single entity when they in fact represent two different enemy vehicles), from deliberately fabricated data being fed into the network or from data that is temporally no longer correct (Bolia et al., 2007). The SA-related information presented by the system should therefore be accurate and free from spurious data at all times.

5. *User trust/confidence in the SA-related information presented by the system.* Trust in other team members and the technology being used is a critical element in the acquisition and maintenance of Distributed Situation Awareness. The users of the system should implicitly trust the SA-related information that is presented to them at all times. Endsley & Jones (1997) suggest that confidence in data is a particular problem in digital MP/BM environments since information is often dated, conflicting, interpreted incorrectly or patently false.

6. *Support for global (or common) SA and compatible SA requirements.* The Distributed Situation Awareness model described demonstrates how Distributed Situation Awareness is built on the basis of transactions in SA between team members and also those different team members have different, but compatible, SA requirements. The nature of mission planning and execution activities in the military land warfare domain is such that in addition to requiring the overall picture (that is, global SA) different agents with different goals and roles require different types of information. Additionally, different agents often use the same information very differently. The system should therefore possess the capability to present both global and compatible SA-related information to its users and the users should be able to easily toggle between the different perspectives. Users should also be able to tailor the system to their own specific needs (that is, be able to customise the interface, tools available, information presented and format of the information presented). Further, the system should support SA transactions by presenting related Distributed Situation Awareness information together and allowing users to determine who needs what information and when.

7. *Support for the level of* Distributed Situation Awareness *that is required to support efficient, timely and effective mission performance.* Ultimately, mission planning and execution support systems should be judged on the overall level of Distributed Situation Awareness that they afford and whether or not they provide the system with sufficient information, and subsequently the appropriate level of Distributed Situation Awareness, required for mission success.

Overall, these questions relate to the main point of our Distributed Situation Awareness assessment, namely whether or not the digital MP/BM system improved or degraded Distributed Situation Awareness during the planning and execution activities that we observed (compared to the original paper map process). The purpose of this analysis was to evaluate the digital system in terms of the requirements specified above. Each requirement formed a sub-hypothesis with the positive outcome as the expected outcome.

Methodology

Design

The study involved a live observational study of an operational field trial of the digital mission planning system in question. The 3-week trial involved a fully functional Division, Brigade (Bde) and Battle Group (BG) undertaking mission planning and execution activities using the new digital system. The trial was set up specifically in order to test the new system and closely represented a real-world operational situation.

Participants

The participants involved in the study were the army staff working in the Bde and BG teams involved in the operational field trial of the mission support system in question. The Bde and BG Headquarters (HQs) analysed consisted of the following cells for the staff to work from; G3 Operations; G5 Plans; G6 Operations; Combat Systems Support Operations (CSSO); Air Aviation; G2 Intelligence; ISTAR; I-Hub; Artillery and Engineers. Due to the nature of the study and data restrictions, it was not possible to collect participant demographic data. A diagram depicting the Bde and BG HQs and the component cells is presented in Figure 6.2

Materials

The materials used for this study included the digital mission support system terminals, the resources used by the Division, Bde and BG throughout the trial, including the materials at each HQ (for example, maps, pens, VHF radios, acetates, tables, chairs, Smartboards, SOIs and so on) and also on the battlefield (for example, vehicles, equipment, weapons and so on). The materials used by the analysts to collect the data included notepads and pens, digital cameras and audio recording equipment.

Procedure

The procedure involved two main components: firstly, an analysis was undertaken in order to identify the Distributed Situation Awareness requirements of the Bde and BG members during the mission planning process and, secondly, an analysis of Distributed Situation Awareness during the mission planning and execution activities observed was undertaken.

For the SA requirements analysis, a HTA was constructed for the planning process using data derived from SOIs and interviews with Subject Matter Experts (SMEs). The HTA description was then refined over the course of the field trial on the basis of observations and further interactions with SMEs. The SA requirements of the different team members involved were then extracted from the HTA and were used to construct propositional networks for each team member involved in the planning process.

For the analysis of Distributed Situation Awareness during mission planning and execution activities, a total of six analysts located within the Bde and BG HQs undertook direct observation of the planning and battle execution activities over the course of the 3-week trial. The analysts were located within the HQs during both planning and execution activities. The data recorded during the observations included a description of the activity (that is, component task steps) being performed by each of the agents involved, transcripts of the communications that occurred between agents during the scenarios, the technology used to mediate communications, the artefacts used to

Figure 6.2   Bde/BG HQ layout showing component cells

aid task performance (for example, tools, computers, instructions, substation diagrams and so on), the temporal aspects of the tasks being undertaken (for example, time undertaken, time available and time taken to perform tasks), and any additional notes relating to the tasks being performed (for example, why the task was being performed, what the outcomes were, errors made, impact of the system on task and so on). Analysts were also given access to planning products, SOIs, logs, briefs and SMEs throughout the field trials. To back up the data collected during the observations the analysts frequently held discussions with the participants and SMEs. Based on the data collected, propositional networks were developed for the mission planning and battle execution activities observed. Propositional networks were constructed for the Bde and BG planning and battle execution activities observed by the HFI DTC analysts during the exercise. The construction of the propositional networks involved conducting content analyses on verbal transcripts, live observation and/or HTA data collected during the observations. Examples of the propositional networks constructed and a summary of the initial findings derived from the propositional networks are presented on the following pages.

Results

Situation Awareness Requirements Analysis

The SA requirements of each of the different team members (Bde and BG) involved were extracted from the HTA that was developed for the planning process. These were then used to develop propositional networks depicting the SA requirements of the different team members during each of the seven questions planning phases. For example purposes, the team member SA requirements for questions one and two are presented in Figures 6.3(a) and 6.3(b).

Seven Questions Planning Analysis

The seven questions planning process (at Bde level in this example) was analysed throughout the exercise using the propositional network approach. A high-level task model of the seven questions process observed is presented in Figure 6.4.

The task model shows, albeit at a high level, the critical activities that the Bde and BG were engaged in during the planning process. These included receiving orders, undertaking the seven questions planning process, conducting a comparison of the COAs developed, the Commander's decision, wargaming the chosen CoA and then producing and delivering the appropriate orders. Alongside these activities were the constant processes of communicating up and down the command chain and within Bde and BG, and also managing information and the various products and outputs derived from the planning process.

Propositional networks were developed for the Bde digital MP/BM-supported seven questions planning process observed during the exercise. The seven questions propositional networks are presented in Figure 6.5 to Figure 6.11.

The propositional networks presented depict the information required during the seven questions planning process. It is notable that, in its present form, the digital system presents the information necessary to support the seven questions planning process and it also provides users with the necessary functionality (that is, tools) required to undertake the seven questions. Despite this, issues associated with the timeliness and accuracy of the information presented, the presentation of appropriate (that is, required) information to users with different SA requirements and the usability of the digital system's planning tools all adversely impacted the level of Distributed Situation Awareness that the BG held during the planning activities observed. For example, the tempo of planning was reduced due to problems with the timeliness of the information being presented. In addition, the users often had doubts regarding the accuracy of the information presented to them by the digital system, which led to them querying the information and undertaking further processes required to clarify the data. This also adversely impacted tempo during planning. The usability issues with the system's planning tools (for example, map displays, user-defined overlays, synch matrix, maps, TASKORG and so on) also impacted the tempo of planning activities, since the users had various problems using them and so took too long to develop planning products. Finally, the lack of support for the different SA requirements of the different cells involved in the planning process (for example, G2, G6, Artillery, Engineer and so on) meant that users had to find and locate the information that they required, which was often time consuming and error prone and served to delay the acquisition of SA. These issues are examined in detail in the discussion section of this chapter.

The key information elements were extracted from the planning execution propositional networks using the five or more links rule. The key information elements are presented in Table 6.1. The key information elements are useful in this case as they represent the key pieces of information that were critical to the planning process. In relation to system design this information is useful since it can be used to strengthen or increase the communication channels that are used to disseminate key information elements within a particular system, or introduce new interfaces that present this information more explicitly.

Question one

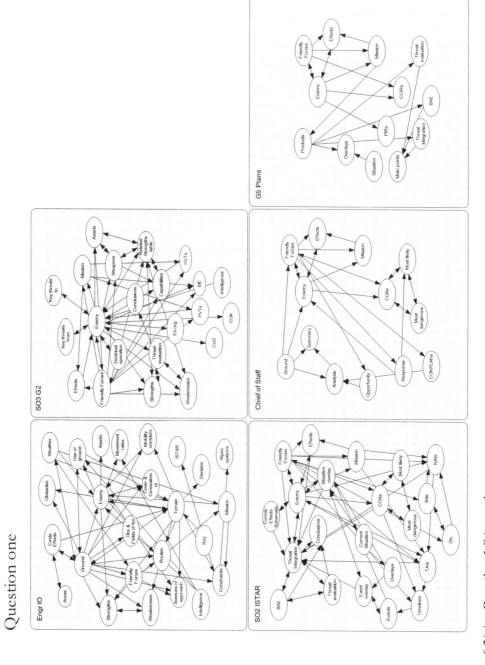

Figure 6.3(a)   Question 1 SA requirements

Question two

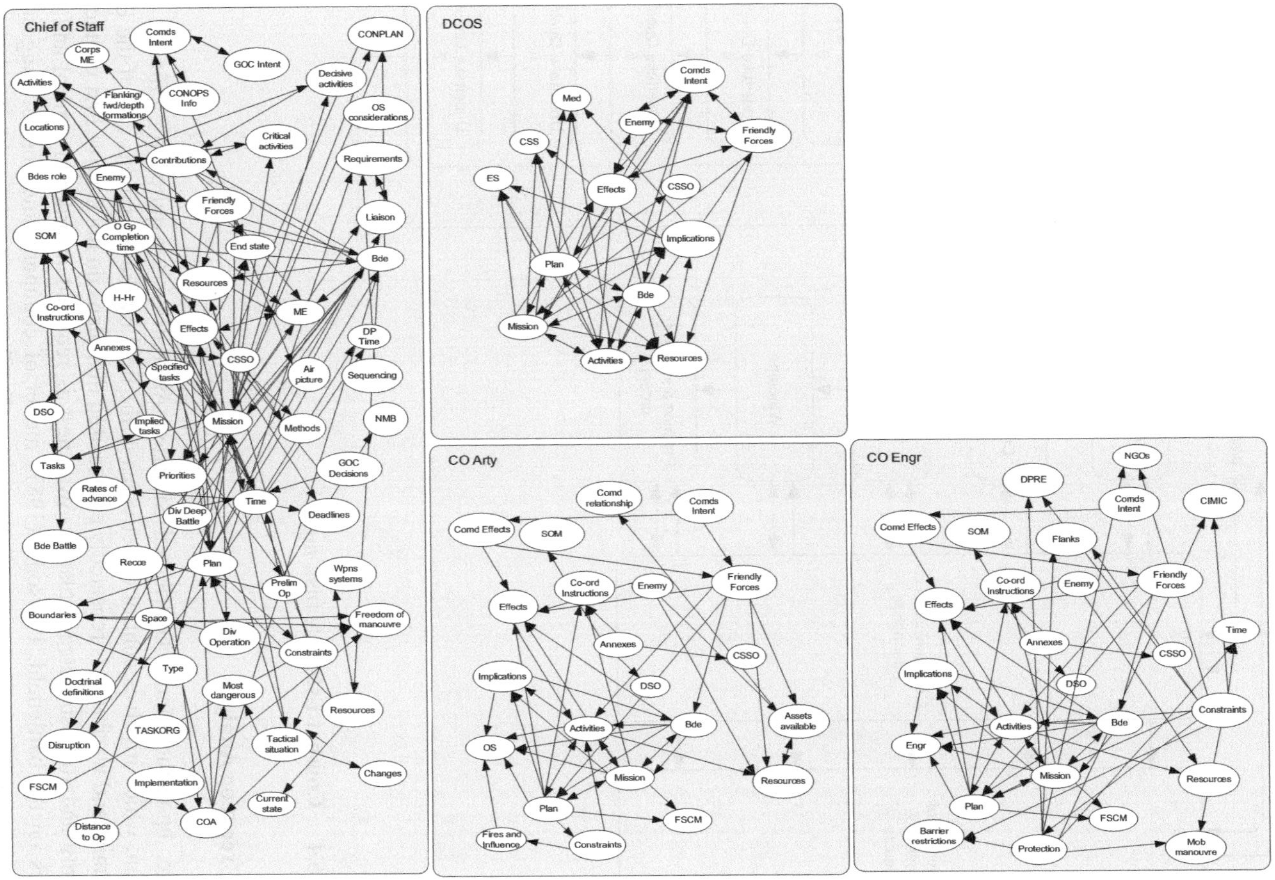

Figure 6.3(b)   Question 2 SA requirements

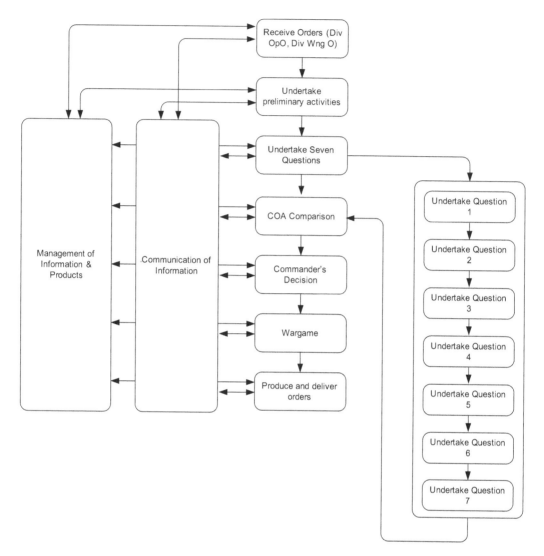

Figure 6.4   Combat Estimate task model

Battle Execution Analysis

Accurate, up-to-the-minute Distributed Situation Awareness is required for efficient battle execution, which involves control of activities in the field by the operations cell of the Bde or BG. Within land warfare it is repeatedly emphasised that plans do not survive first contact with the enemy and so Distributed Situation Awareness is required in order to control and direct activities on the battlefield. Lawson's (1981) model of command and control suggests that data are extracted from the environment, processed and then compared with the desired end state. Discrepancies between the current state and the desired end state serve to drive decisions about how to move towards the desired end state once more. These decisions are then turned into actions and communicated to the forces in the field. Distributed Situation Awareness is

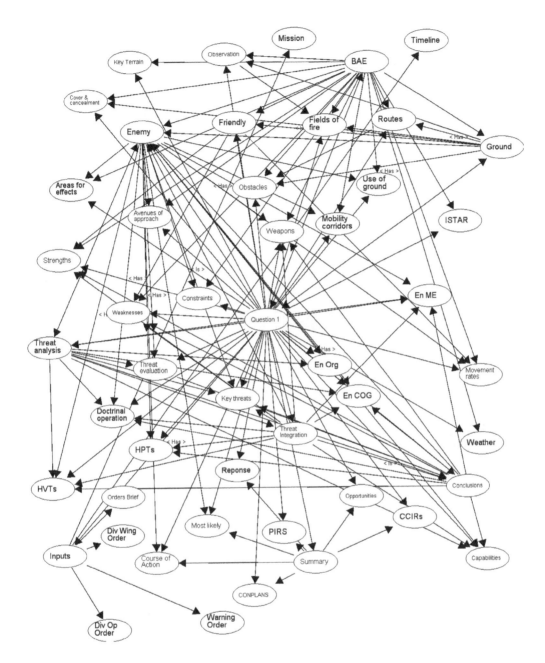

Figure 6.5 Question one propositional network

obviously of critical importance here, since it is the accurate understanding of the current situation that is compared to the desired end state. Of primary importance then during battle execution are the provision of an accurate and up-to-date Local Operational Picture (LOP), and also the presence of communications links for facilitating Distributed Situation Awareness acquisition and maintenance. For the purposes of this book a Distributed Situation Awareness-

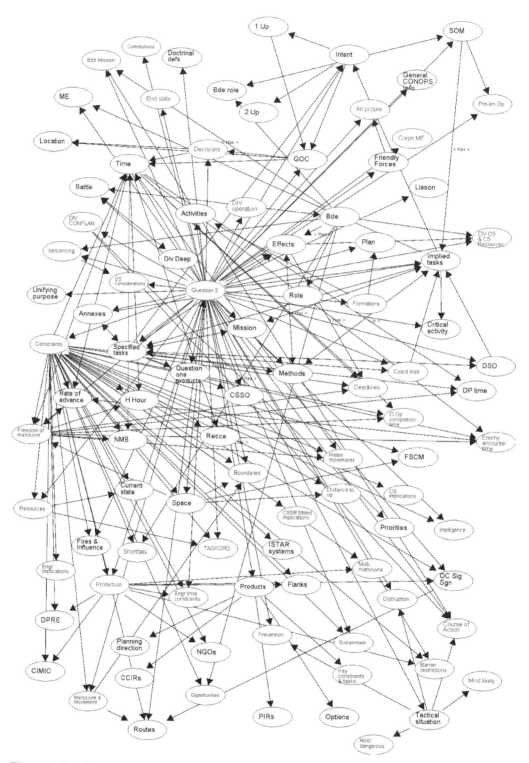

Figure 6.6   Question two propositional network

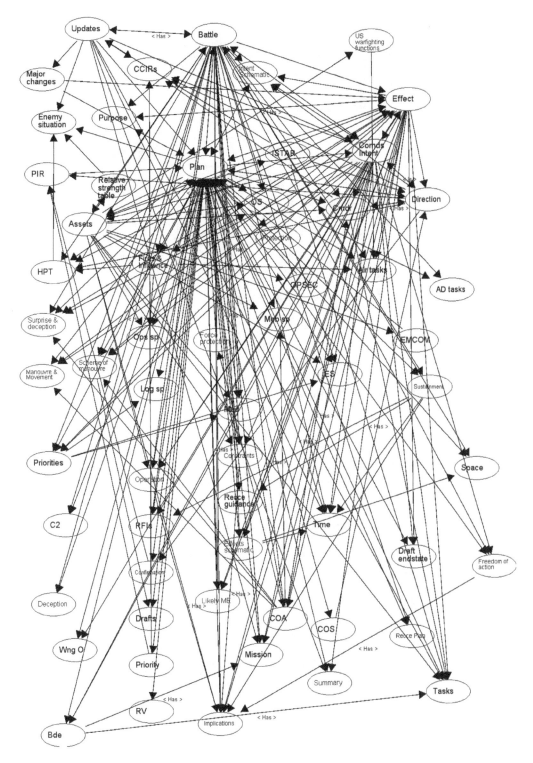

Figure 6.7   Question three propositional network

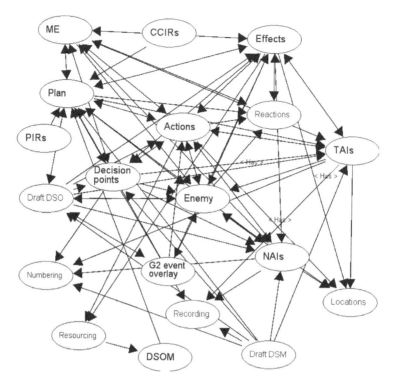

Figure 6.8   Question four propositional network

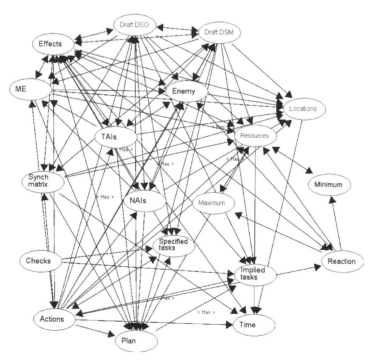

Figure 6.9   Question five propositional network

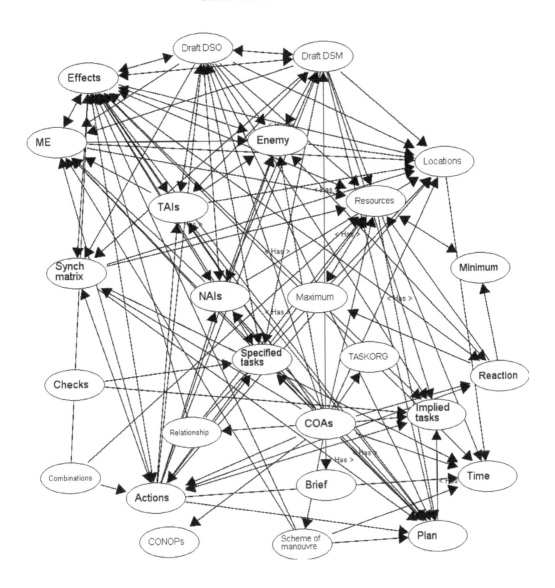

Figure 6.10 Question six propositional network

based analysis of the battle execution activities undertaken by the BG during battle two (battle two occurred on Friday 16th November) is presented. Propositional networks were constructed for the battle execution activities based on content analyses of the verbal communications taking place between the key agents located at the ops table in the BG. The ops table layout is presented in Figure 6.12.

A task model was constructed for the battle execution activities observed. Task models provide a high-level representation of the key tasks involved. The task model for the battle is presented in Figure 6.13.

The task model shows, albeit at a high level, the critical activities that the BG was engaged in during the battle. These included developing and maintaining an accurate battle picture,

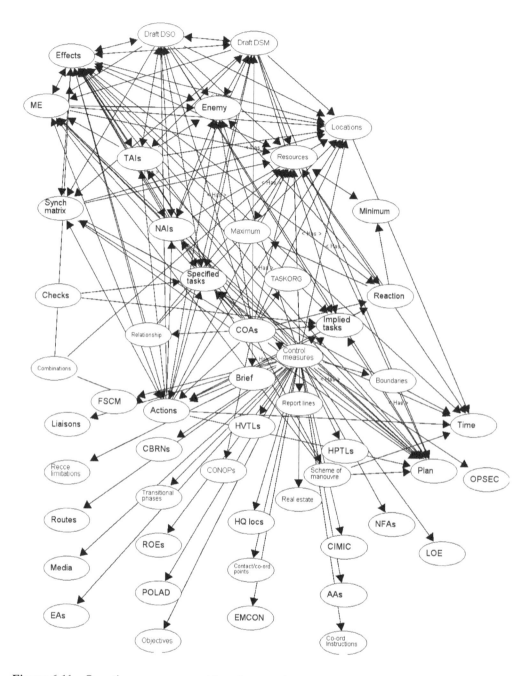

Figure 6.11   Question seven propositional network

developing and distributing orders, coordinating and directing the enactment of the plan, and communicating with assets and other elements of the command chain (via situation reports and so on). It is noteworthy that each of these activities is interlinked and is ultimately reliant on the system providing the BG with a highly accurate and timely Distributed Situation Awareness. For example, without accurate Distributed Situation Awareness of what is happening on the battlefield

Table 6.1   Planning key information elements

KEY INFORMATION ELEMENTS	Question One		Question Two		Question Three		Question Four		Question Five		Question Six		Question Seven	
	BAE	Enemy	Intent	Implied tasks	Mission	Implications	Effects	ME	Draft DSO	Effects	Draft DSO	Effects	Control Measures	Effects
	Ground	Strengths	Specified tasks	Question 2	Cmdrs Intent	Effects	Plan	TAIs	Draft DSM	TAIs	Draft DSM	TAIs	Draft DSO	TAIs
	Weaknesses	Threat analysis	Time	Effects	Plan	Direction	NAIs	Locations	NAIs	Locations	NAIs	Locations	Draft DSM	Locations
	Threat Integration	Question 1	Mission	Constraints	Battle	Updates	Actions	DPs	Actions	Resources	Actions	Resources	NAIs	Resources
	Weapons	Key threats	FOM	Rates of advance	CCIRs	Assets	Draft DSO	Enemy	Draft DSO	ME	Draft DSO	ME	Actions	ME
	Enemy COAs	Capabilities	Space	Protection	Fires & Influence	Cmdr			Synch Matrix	Reactions	Synch Matrix	COAs	Draft DSO	COAs
	HVTs	Inputs	Tactical situation	Products	OS	Main tasks			Implied tasks	Specified tasks	Implied tasks	Specified tasks	Synch Matrix	Specified tasks
	Summary	Mobility Corridors	Effects	Deadlines	SOM	Time			Time	Plan	Time	Plan	Implied tasks	Plan
	Conclusions	Enemy Org			Tasks	Implications			Enemy		Enemy	SOM	Time	SOM
	Enemy COG				Intent schematic	HPTs							Enemy	
					Priorities	Surprise & Deception								
					Sustainment	COA								

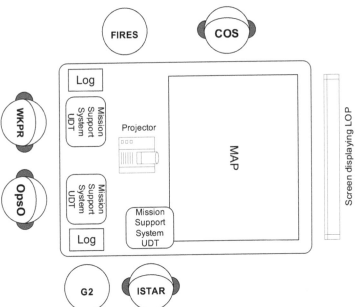

Figure 6.12   Ops table layout

the BG cannot develop and maintain an accurate battle picture and thus cannot develop appropriate orders or direct the plan enactment. The provision of accurate situation reports up, down and within is also dependent upon them having an accurate picture.

Distributed Situation Awareness during battle two was analysed using the propositional network approach. A total of six propositional networks were constructed, for the activities occurring during enactment of each of the phase lines involved in the friendly force plan (Harry and Scabbers,

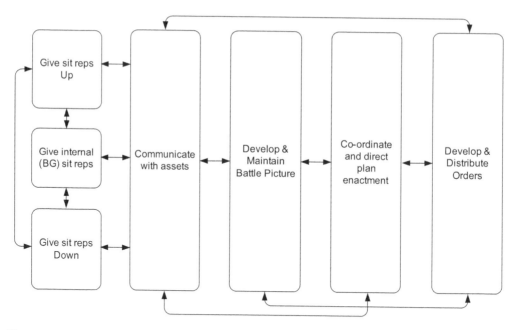

Figure 6.13   Battle Execution task model

Voldemort, Hagrid, Dobby, Dumbledore and Hedwig). The propositional networks are presented in Figure 6.14 to Figure 6.19.

The key information elements were extracted from the battle execution propositional networks using the five or more links rule. The key information elements are presented in Table 6.2.

The shaded items within Table 6.2 represent those information elements that were constant throughout the battle (that is, were key information elements during each phase line enactment). The key information elements are useful in this case as they represent the key pieces of information that were critical to the battle execution process. In relation to system design this information is useful since it can be used to strengthen or increase the communication channels that are used to disseminate key information elements within a particular system, or introduce new interfaces that present this information more explicitly. In this case, it is notable that there were issues surrounding the accuracy and presentation of a number of the key information elements presented in Table 6.2. For example, enemy and friendly force location information was typically not presented in a timely fashion and so was often not compatible with the actual state of the world.

The propositional networks in this case are useful not only because they depict the information used by the BG during enactment of the battle, but also because they contain within them naturalistic examples of problems encountered with the digital MP/BM system, and also instances in which the digital MP/BM system significantly impacted the processes involved. A summary of the different issues observed during planning and enactment of the battle is presented overleaf.

Discussion

The main aims of the study were to analyse Distributed Situation Awareness during land warfare mission planning and execution activities and to evaluate the new digital system in terms of its

Table 6.2   Battle execution key information elements; (shaded items denotes information elements that were transacted across phase lines battle execution process)

KEY INFORMATION ELEMENTS	Harry & Scabbers		Voldemort		Hagrid		Dobby		Dumbledore		Hedwig	
	Contacts	Picture	Battle	Picture	Friendly Forces	Picture	Contacts	Picture	Enemy	Picture	Enemy	Picture
	Friendly Forces	Enemy	Friendly Forces	Enemy	Plan	Enemy	Friendly Forces	Enemy	Plan	Battle	Plan	Battle
	Plan	Situation	Locations	Situation	Voldemort	Situation	Plan	Situation	Voldemort	Assets	Phase lines	Assets
	Map	Locations	Opening	Time	Assets	Locations	Battle	Assets	Phase lines	Situation	Friendly Forces	Situation
	Op Order	Icons	Phase lines	Javelin	Phase lines	Battle	Time	Phase lines	Time	Friendly Forces	Time	Cancellation
	Obstacles	TAIs			Cancellation	Time	Marking up	Locations	Locations	Sit reports	Grid refs	Locations
	NAIs	Minefield				Javelin	POWs	Sit reports	Marking up	Likely enemy positions	BMDs	Comms
	Grid refs	Call signs						Kills	FOO	Dobby	Sit reports	Contacts
	Crossing site	Assets							Grid refs	Contacts	Scabbers	
	Bridges	Time							POWs	Kills		
	Phase lines	Battle							River	Crossing		
									Intent	Remaining Platoon		
									Platoons	Recce vehicles		

impact on Distributed Situation Awareness during the activities observed and its support for Distributed Situation Awareness requirements.

Distributed Situation Awareness During Mission Planning and Execution

The analysis revealed a number of interesting facets associated with Distributed Situation Awareness during the mission planning and execution activities analysed. From a theoretical viewpoint, it was notable that the notion of compatible, rather than shared, SA was apparent during the planning and battle execution activities. The propositional networks indicate that the information elements underlying the systems Distributed Situation Awareness represented compatible, rather than shared or common SA requirements. Each team member was using this information for their own means and the distinct roles and responsibilities extant throughout the planning and battle execution process were such that common or shared SA was neither possible nor would it have been productive. This finding was corroborated by the SA requirements analysis findings, which suggested that each team member had distinct SA requirements. The corollary of this was that, even when different staff were using the same information, they were using it for different purposes and so their SA was different to one another.

During the planning process, the planning team is divided into distinct cells, each with their own specific role and subsequent goals and tasks to fulfil. For example, during question one (the Battlefield Area Evaluation (BAE) phase), the engineer cell is primarily concerned with the ground, friendly and enemy forces use of the ground and the impact that the ground is likely to have on friendly and enemy operations, whereas the intelligence cell is primarily concerned with the threat posed by the enemy, including the enemy's capability, strengths and

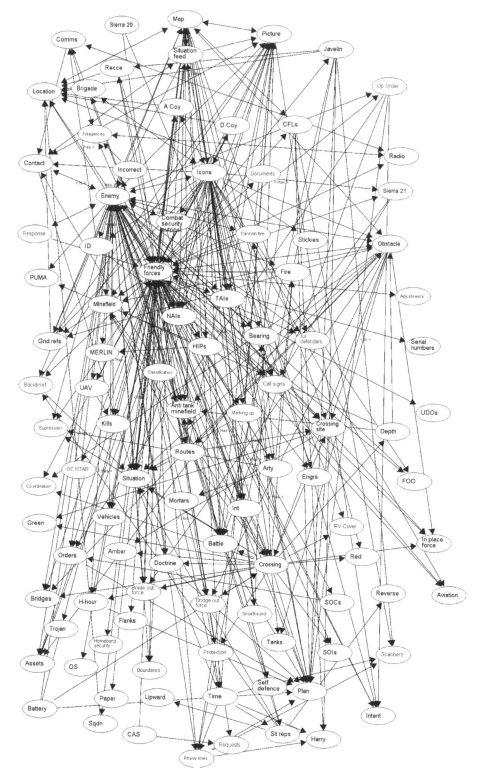

Figure 6.14   Harry & Scabbers propositional network

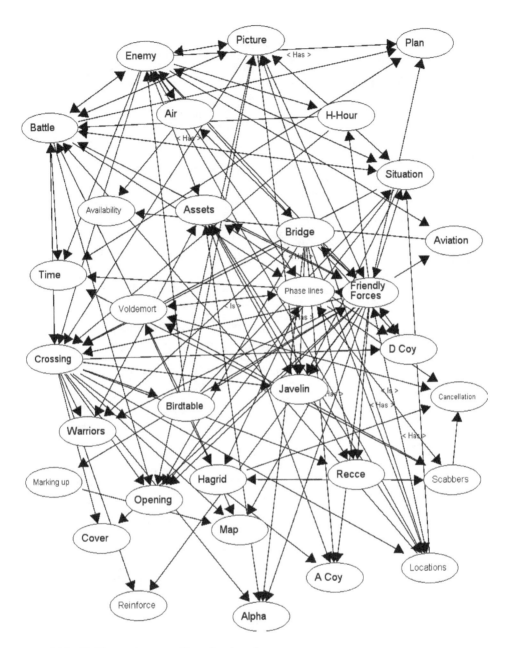

Figure 6.15   Voldemort propositional network

weaknesses, and enemy doctrine. Thus even when both team members have access to the same information regarding the battlefield area and the enemy, they use and view the information in a very different manner; it is the relationship between concepts that makes up their distinct SA. Indeed, thinking about SA as the relationship between concepts is the key to the Distributed Situation Awareness approach; even when team members have access to the same information, the relationships between the information elements is likely to be different based on how they are using the information and what they need it for. In the example cited above, the relationships

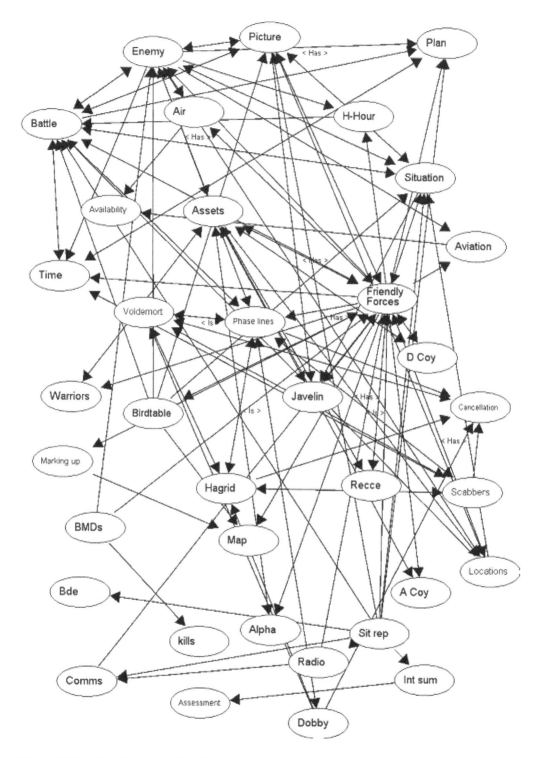

Figure 6.16   Hagrid propositional network

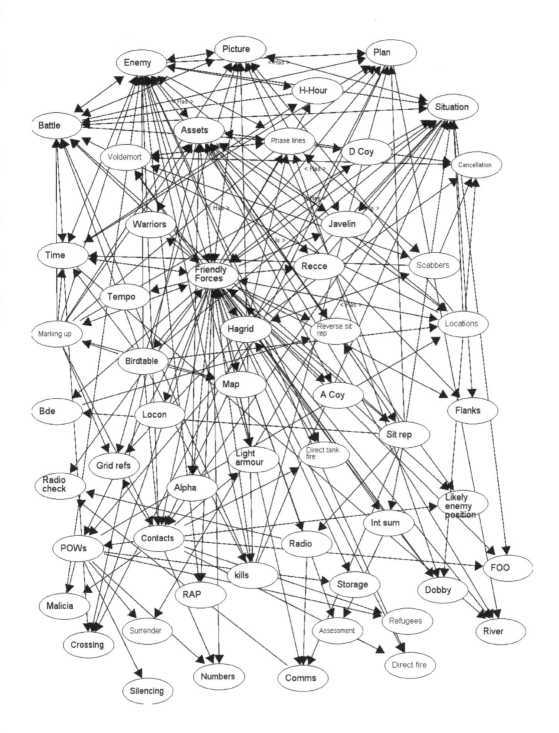

Figure 6.17 Dobby propositional network

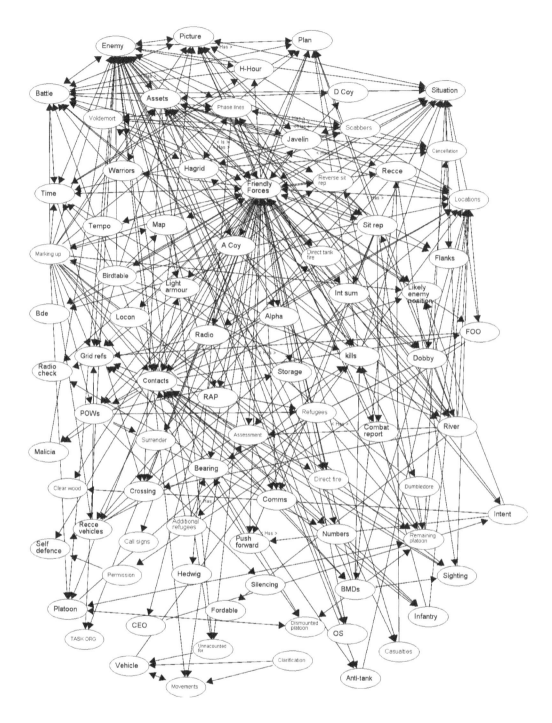

Figure 6.18   Dumbledore propositional network

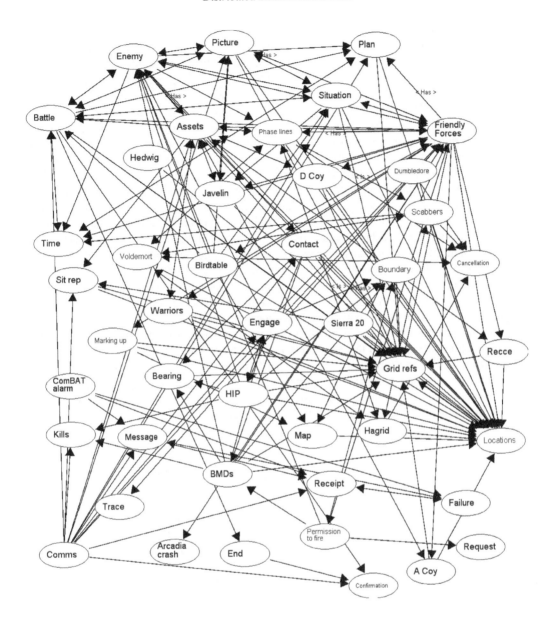

Figure 6.19 Hedwig propositional network

between the enemy and the battlefield area are viewed very differently by the engineer and the intelligence components; the engineer looks at how the ground may shape enemy operations whereas the intelligence cell looks at the ground and the resultant threat level imposed by the enemy. It is this unique combination of information elements by each team member that makes their SA compatible and not shared; the very fact that an actor has received information, acted on it, combined it with other information and then passed it on to other actors means that its interpretation changes per team member; this represents the transaction in awareness referred to earlier. The engineer versus intelligence components views on the enemy and ground during question one is represented in Figure 6.20.

This clear distinction between roles and responsibilities brings with it distinct SA requirements for each cell. It is therefore concluded that the organisation of the teams involved and the presence of specialised roles was such that the majority of the SA elements represented distinct compatible SA elements; each individual and sub-team had had their own unique combination of Information Requirements (IRs) depending on the role, goals and tasks that they were required to undertake.

This compatibility of team member SA requirements underlying Distributed Situation Awareness during the planning and battle execution activities observed has very clear implications for the design of any system intended to support them (and for collaborative systems in which team member roles are distinct). The implication of this is that collaborative system design should be driven with a very clear specification of the compatible SA requirements of the different users; the system should then be tailored to support these unique SA requirements. This permits a system that has the capability to present only the required information to the right users at the right time, a provision that is key for Distributed Situation Awareness. Rather than simply design a system that presents all of the information available (requiring users to locate the information that they require) and contains all of the planning tools and functions required by the overall group (requiring users to locate the functions that they require), the tool should instead be tailored specifically to support each role (in terms of information and tools required). User-tailored systems such as this would minimise the overload of agents with unwanted information and tools.

This conclusion is corroborated by other findings presented in the literature. For example, Bolstad, et al. (2002) analysed the SA requirements of a US Army Bde and also found explicit differences between the SA requirements of the different officers. In conclusion, they suggested that, in military planning systems, team members do not need to know everything that the other team members know; this meant that a single display would not meet the needs of all of the Bde officers. Subsequently Bolstad et al. (2002) recommended that in order to provide only the level

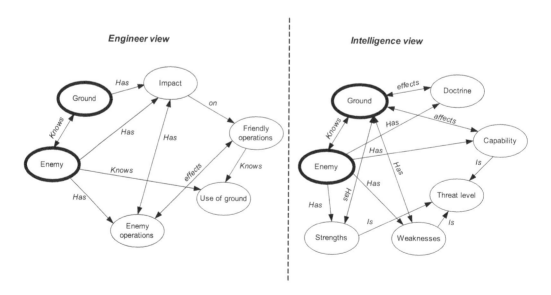

Figure 6.20   Engineer versus intelligence components differing views on enemy and ground information elements; figure demonstrates how each component is using the information for their own ends and how its subsequent combination with other information makes each view unique

of detail required for a particular user, without presenting unnecessary information, displays should be tailored to each officer's needs whilst also providing information relating to the SA of the other officers in the team. Gorman et al. (2006) also suggested that, due to the specialised roles apparent within typical command and control environments, the design principle of giving every team member displays which present all of the information required by the entire team is invalid. Gorman et al. (2006) proposed that it may in fact be prohibitive and counteractive to give everyone mutual access to the same information. Similarly, Kuper & Giuerelli (2007) postulate that in order to enhance command and control team efficiency, tailored work aids should be used to reduce the cognitive load associated with mining through redundant information. They argued that the key to efficient and effective command and control team performance is the design of work aids that support both holistic work practices and unique first person perspectives.

The present analysis indicated that, despite the presence of such explicitly compatible, rather than shared, SA requirements, this has not been taken into account in the design of the digital MP/BM system. In its current form, the digital MP/BM system does not support the compatible SA requirements of its different users. Rather, the system simply provides the same displays, tools, interface and, more importantly, information to every user regardless of roles and goals. It is not customisable nor can it be tailored based on different user requirements. The onus is thus placed on the user to find the information and tools that they need within the system, a process which ostensibly is time consuming and difficult.

Digital MP/BM Systems Impact on Distributed Situation Awareness

The analysis also provided compelling evidence of the impact on Distributed Situation Awareness that the digital MP/BM system had during the activities observed; these findings can be discussed with regard to the hypotheses set out at the beginning of this chapter. The first and perhaps most telling finding was that, in undertaking the required activities, the teams involved continued to use the traditional paper map processes in order to support and supplement the new digitally supported process. On the majority of occasions, this was because of flaws present within the digital MP/BM system that were adversely affecting Distributed Situation Awareness. It was simply quicker and easier for users to generate and maintain the level of Distributed Situation Awareness required using the old paper processes. It was concluded that the digital MP/BM system did not adequately support the acquisition and maintenance of Distributed Situation Awareness during the activities observed; rather, a combination of the paper map process and the digital MP/BM system was used.

Secondly, there were many instances in which the SA-related information presented by the digital MP/BM system was in fact inaccurate and was not compatible with the real state of the world at the time when it was presented. This is represented in the summary propositional network presented in Figure 6.21, which depicts a summary of the planning and execution information elements used. Within Figure 6.21, the information elements that were presented inaccurately by the digital MP/BM system are shaded as dark grey.

This was particularly problematic during battle execution, where the information presented on the LOP was either out of date or spurious. This meant that the Bde and BG's understanding of enemy and friendly force locations, movements, number and capabilities was often inaccurate. To overcome this, radio voice communications were used to supplement and/or clarify contact reports and a paper map with sticky icons was used to represent the battle. These mismatches had the impact of reducing the accuracy of Distributed Situation Awareness and also adding time to the planning and execution; the result of this was a reduction (rather than the projected increase) in operational tempo.

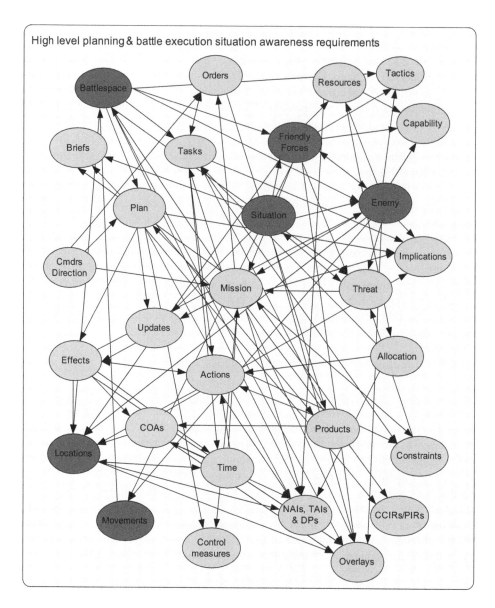

Figure 6.21   Inaccurate information elements; those nodes shaded dark grey represent the information that was presented inaccurately by the digital MP/BM system

Thirdly, the timeliness of the SA-related information presented by the digital MP/BM system was also problematic; due to data bandwidth limitations voice transmission was given precedence over global positioning data regarding the locations and movements of entities on the battlefield. Because of this, contact reports and positional information presented on the LOP was often up to 20 minutes late. This is represented in Figure 6.22.

The corollary of this was that the system's Distributed Situation Awareness at times appeared to be 'out of date' or at least lagging behind the real state of the world. Ostensibly, the problem of 'delayed' SA information presented by the system was a bandwidth issue. Specifically, because of the amount of data being transmitted and the limited bandwidth of the system, the voice communications

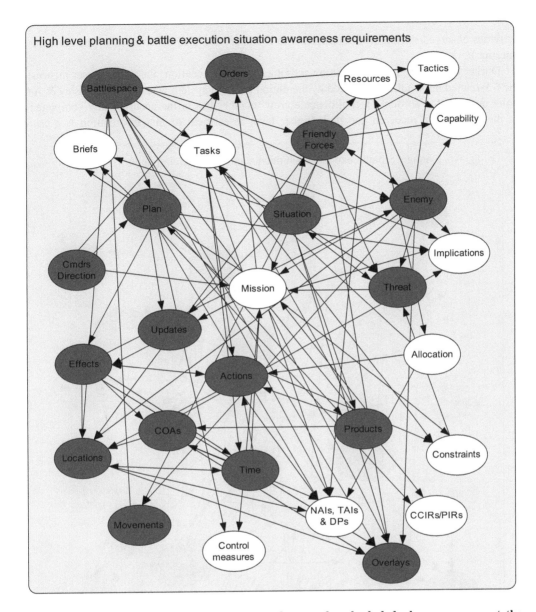

Figure 6.22 **Untimely information elements; those nodes shaded dark grey represent the information that was not presented in a timely manner by the digital MP/BM system**

data takes precedence over the Own Situation Position Report (OSPR) data. This meant that during complex operations the OSPR data is delayed due to high voice communications traffic. Due to the same data transmission problems the digital system was also observed to be slow in updating the enemy positions on the LOP.

As a consequence of the problems discussed above, a fourth issue identified was the low level of trust that the users placed in the SA-related information presented to them by the digital MP/BM system. This is represented in Figure 6.23, where the information elements shaded in dark grey represent the information that users of the digital MP/BM system did not fully trust during the activities observed (based on discussions with the users and also observation of the activities in question).

During both planning and execution activities (mainly execution) the issue of user mistrust in the SA-related information presented by the digital MP/BM system was evident. Endsley & Jones point out that 'the amount of confidence a crew member has in the accuracy and completeness of the information received and their higher level assessment of that information is a critical

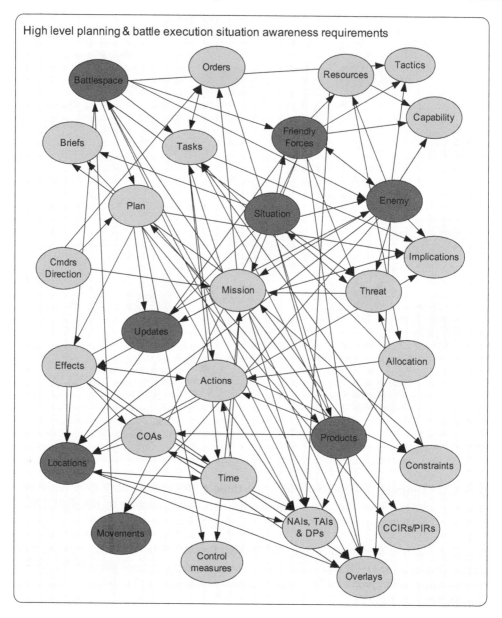

Figure 6.23   Lack of trust in information elements; those information elements shaded in dark grey represent that information presented by the digital MP/BM system that users felt was untrustworthy

element of situation awareness' (Endsley & Jones, 1997, p. 28). According to Endsley & Jones, in the event of uncertain information, individuals either search for more information or act on uncertain information, both of which can be detrimental to SA. It is apparent from the analyses that user trust in the information presented by the digital MP/BM system was minimal; due to issues such as SA mismatches, spurious data and significant delays in the presentation of positional and contact reports many users often questioned the information presented by the digital MP/BM system and often took measures to clarify the accuracy of the information (for example, requests for clarification of location and status reports). This served to add to the planning and execution process and also adversely impacted the tempo of operations.

A fifth and final issue related to the granularity of the maps used (within the digital MP/BM system) and their impact on Distributed Situation Awareness was identified. One of the key issues related to the development of SA of the ground during the planning process that was observed consistently throughout the exercise was the problems with the granularity of the maps used on the digital MP/BM system. Users found it extremely difficult to analyse the ground and appreciate what was going on when looking at the maps presented by digital MP/BM system. This meant that the users could not assess the ground sufficiently and, in addition, the size of the display meant that users could not get an overview of the entire battlefield. The only way in which users could see the entire battlefield area was to zoom out, however, this often led to the users losing context in terms of exactly which area of the battlefield area they looking at. This problem was consistently reported by users throughout the exercise.

Overall, the analysis suggests that the digital MP/BM system did not provide adequate support for Distributed Situation Awareness development and maintenance during the planning and execution activities observed. Rather, it was a combination of the digital MP/BM system and the traditional paper map process that enabled the system to develop and maintain the level of Distributed Situation Awareness required for successful completion of planning and execution activities.

Conclusions

In closing it is our opinion that, in isolation, the digital MP/BM system analysed did not provide adequate support for Distributed Situation Awareness during planning and execution activities. Rather, it was a combination of the digital MP/BM system and the analogue paper map process that enabled the system to develop and maintain the level of Distributed Situation Awareness required for successful completion of the planning and execution activities observed. Although the digital MP/BM system does appear to have the potential to support Distributed Situation Awareness during planning and execution activities, at present it falls short of this key requirement in a number of areas and consequently, a combination of both planning systems (digital and analogue) was used throughout the activities observed. The issues limiting the level of Distributed Situation Awareness afforded by the new digital MP/BM system included the timeliness and accuracy of the information presented and the presentation of appropriate information to the appropriate users, all of which subsequently impact the trust that the users place in the SA-related information presented. On a positive note, the digital MP/BM system does appear to present the information required for Distributed Situation Awareness and also provides the communication links required for Distributed Situation Awareness to percolate throughout the system.

The huge potential of digitising warfare systems and processes can only be realised with further investigation and evaluation in order to determine how systems can be better designed in order to enhance Distributed Situation Awareness and ultimately mission planning and execution activities.

Key issues to pursue relating to the concept of SA include what information should be presented, in what manner and to which elements of the warfare system, how information can be presented in a more timely fashion and how the accuracy of information presented by command and control systems can be enhanced and ensured. Ultimately the great potential that digitisation offers for enhancing mission planning and execution activities in the military domain is also accompanied by a very real opportunity to create warfare systems in which activities become more difficult and complex, more prone to error and subsequently less efficient.

Chapter 7
Social Network Analysis

Aim of the Chapter

This chapter takes the NATO SAS-050 Approach Space, a widely accepted model of command and control, and gives each of its primary axes a quantitative measure using Social Network Analysis (SNA). Deriving such measures means that the actual point in the approach space adopted by real-life command and control organisations can be plotted, along with the way in which that point varies over time and function. Part 1 of the chapter presents the rationale behind this innovation and how it was subject to verification using theoretical data. Part 2 shows how the enhanced approach space was put to use in the context of a large-scale military Command Post Exercise (CPX). Issues regarding emergent properties and interface bottlenecks were revealed by the analysis, which was further extended to offer quantitative insights into agility and tempo. The main findings are based on a comparison between where the organisation thought it was in the approach space, where it actually was, and the extent to which there was a mismatch between it and the problem to which it was directed. Above all, the findings show that it was the humans in this particular live Networked Enabled Capability Network Centric Warfare (NEC) situation that granted the levels of agility and tempo that were observed.

Introduction

Background and Context

Many visions exist as to what NEC should be. They range from something akin to a computer network at one end of the scale (the example of Wal Mart's vertically integrated systems seems to have been an early inspiration; Shachtman, 2007), to an organisation that exhibits the behaviour of an organism at the other (in which the inspiration derives, albeit indirectly, from the legacy of socio-technical systems; Walker et al., 2009). Unsurprisingly, the practical realisation of NEC also varies considerably, with widely different forms of organisational infrastructure (some of it ostensibly 'technical' in nature) combined with operational procedures that vary between the scripting of tasks and micro-management through to a focus on semi-autonomous groups and so-called 'effects-based operations'. As such, the state-of-the-art in terms of the real-world implementation of NEC is currently, and perhaps will always remain, something that is much more akin to a process than it is to a fixed end-state (Alberts & Hayes, 2003).

The NATO C2 Approach Space

In its most generic sense, command and control (also known as C2) is simply the management infrastructure for any large, complex and dynamic resource system (Harris & White, 1987). As the description above testifies, not all 'management infrastructures are alike', and because of this the label 'command and control' may not in fact be very helpful (for example, Alberts, 2007). The NATO SAS-050 Approach Space shows this to be the case quite clearly. When the formal definitions of command and control are brought to bear, as in the doctrinally defined, hierarchical, 'classic C2' sense (for example, Mauer, 1994), the management infrastructure that results occupies only a relatively small area of the approach space compared to the various incipient states of NEC described above, which are widely scattered. The NATO SAS-050 Approach Space, therefore, represents an important reality of C2 and is a well accepted basis for exploring and investigating it (for example, NATO, 2006; Alberts & Hayes, 2006; Alberts, 2007).

The development of the model is reasonably well rehearsed (for example, see NATO, 2006 and Alberts & Hayes, 2006). In summary, it derives from an underlying information processing view of command and control expressed as a generic C2 'Approach' (a cyclical pattern of information collection, sense making and actions). This in turn enabled a comprehensive reference model to be developed, comprised of over 300 variables that map on to the C2 approach. These variables were drawn from fields as diverse as general systems theory, Human Factors, cognitive psychology and operational research (amongst many others). The 300 variables are themselves connected by over 3,000 links and various methods were used, such as UML and MYSQL, in order to undertake a process of 'dimension reduction' to arrive at what are termed 'three key factors that define the essence of [command and control]' (Alberts & Hayes, 2006, p. 74). These are as follows:

x = allocation of decision rights (from unitary to peer-to-peer);
y = patterns of interaction (from fully hierarchical to fully distributed);
z = distribution of information (from tight control to broad dissemination).

The three key factors form intersecting x, y and z axes and the three dimensional (3D) NATO C2 Approach Space proper (an example is shown in Figure 7.1). In theory, a command and control organisation can be positioned along its respective x, y and z axes and its position in the approach space can be fixed. It is possible to go further than this. Because NEC is more akin to a process than it is to a fixed point, which is to say that it behaves as a dynamical system, the changing values of x, y and z over time give rise to successive points in the three-dimensional space. When linked together they form a trajectory, a representation of motion that describes the dynamic behaviour of a complex system like NEC. Within the field of complex systems research the practice of dimension reduction and the creation of a simple coordinate space into which a real-life dynamical system can be plotted is referred to not as an 'approach space' but a 'phase space' (Gleick, 1987). The more extensive the motion within the phase space, the greater the organisation's 'variety' is said to be, and thus, according to the Law of Requisite Variety, its ability to cope with complexity (Ashby, 1956). In the language of C2, as Alberts and Hayes (2006) point out, the greater the number of points in the approach space that can be occupied, then the greater the command and control organisation's agility (Alberts & Hayes, 2006).

What drives this motion through the approach/phase space? What causes command and control organisations to change in terms of the three fundamental dimensions? According to NATO (2006) the two proximal sources of these dynamics are function (in that different parts of a command and control organisation can be configured differently) and time (configured differently at different points in an evolving situation). Both of these dynamics, in turn, are dictated by the distal source of dynamics represented by the causal texture of the environment or problem that the command and control organisation is operating under (and which in turn it is influencing; Emery & Trist, 1978). The NATO approach space is therefore joined by a corresponding 'Problem Space'. This too has three dimensions that form intersecting axes and a 3D space. These are as follows:

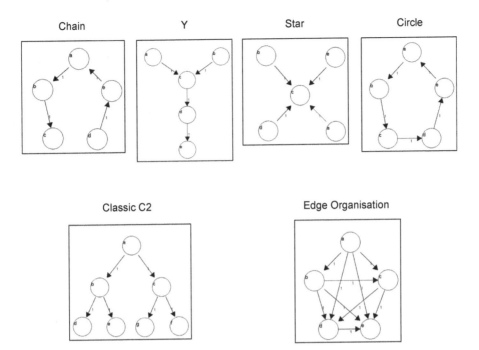

Figure 7.1 **Illustration of archetypal networks. Associated with each is empirical evidence concerning its performance on simple and complex tasks**

x = familiarity (from high to low);
y = rate of change (from static to dynamic);
z = strength of information position (from strong to weak).

Problems that are merely 'complicated' can be characterised by high familiarity (of underlying principles), non-dynamic rates of change (the situation is stable) and consequently a strong information position. The type of problem to which NEC is a conceptual response is not merely complicated but 'complex'. The Systems Engineering phrase is 'a wicked problem', that is, problems that have fuzzy boundaries, lots of stakeholders and lots of constraints with no clear solution (Rittel & Webber, 1973). Complex problems are in turn the ones that can be characterised by unfamiliarity, change and a weak information position, the ones 'that persist – the problems that bounce back and continue to haunt us' (Bar Yam, 2004, p. 14).

The approach and problem spaces are yoked so that the area occupied by a putative 'problem' in one needs to be matched by an organisation occupying a corresponding area of the 'approach' in the other. By these means the variety or degrees of freedom offered by the problem are able to be matched by the variety and/or agility inherent in the command and control organisation (Bar Yam, 2004; Alberts & Hayes, 2006). By fixing and tracking different approaches to command and control, as well as the problems to which they are directed, the ultimate aim of the approach and problem spaces can be met, which is to facilitate exploration of 'new, network enabled approaches […] to command and control and compare their characteristics, performance, effectiveness, and agility to traditional approaches to command and control' (NATO, 2006, p. 7).

The Missing Links

The NATO SAS-050 Approach Space 'is intended to serve as a point of departure for researchers, analysts, and experimenters engaging in C2-related research, conducting analyses of C2 concepts and capabilities, and designing and conducting experiments' (NATO, 2006, p. 3). A number of extensions to the model can, therefore, be usefully and legitimately directed into the following three areas:

1. 'We are interested in the actual place or region in this [approach] space where an organisation operates, not where they think they are or where they formally place themselves' (Alberts & Hayes, 2006, p. 75). The requirement that flows out of this is for metrics to define, quantitatively, the position that live command and control organisations adopt on any one of the approach space's three axes. If live command and control can be fixed into the approach space, then it can be compared 'to traditional approaches to command and control' (or indeed to any other instantiation).

2. The observation that 'an organisation's location in the C2 approach space usually ranges across both function and time' (Alberts & Hayes, 2006, p. 76) brings a further requirement to capture and understand the underlying dynamics of NEC. If the underlying dynamics can be captured and understood, then useful aspects of tempo and agility can be revealed.

3. Finally, 'identifying the crucial elements of the problem space and matching regions in this space to regions in the C2 approach is a high priority'. Fixing and understanding the dynamics of command and control in the approach space increases the accuracy of the mapping that can occur between approach and problem, thus moving judgments about efficacy and performance from relative towards absolute.

Part 1 of this chapter deals with the innovations that enable the NATO Approach Space to be transformed into a phase space in order to meet the three objectives above. An explicit strategy for achieving this is derived from SNA and is put to the test with theoretical data. Testing the hybrid social network/NATO Approach Space with live data in Part 2 provides an opportunity to observe 'the actual place or region where an organisation operates', how that location varies 'according to function and time' and, aided by theoretical data, 'match regions of the problem space to the approach space'.

Part 1: Developing the NATO SAS-050 Model

Social Network Analysis

At the heart of all the missing links presented above is the ability to provide quantitative measures or metrics that relate meaningfully to decision rights, patterns of interaction and dissemination of information. SNA is used to overcome this significant limitation.

In general terms a social network is 'a set of entities and actors [...] who have some type of relationship with one another'. With SNA representing 'a method for analysing relationships between social entities' (Driskell & Mullen, 2005, pp. 58-1–58-6). A social network is created by plotting who is communicating with whom on a grid-like matrix. The entries into this grid denote the presence, direction and frequency of a communication link. The matrix can be populated using information drawn from organisation charts and standard operating procedures so that it describes where an organisation formally places itself. Much more consistent with the approach space is that the matrix can be populated with live data that describes where an organisation is 'actually' placed.

The matrix of agents and links is what enables a social network diagram to be created. This is a graphical representation of the entities and actors who are linked together and obviously, apart from very simplistic networks, any underlying patterns extant in this graphical representation are difficult to discern by eye alone. Thus, graph theoretic methods are applied to the matrix in order to derive a number of specific social network metrics (for example, Harary, 1994). These form the basis of a comprehensive diagnosis of the network's underlying properties, which include several that relate to decision rights, patterns of interaction and distribution of information. This mapping of social network metrics to the NATO model axes is a key innovation and is described below:

Decision Rights Mapped to Sociometric Status

Decision rights can be mapped to the social network metric called 'sociometric status', which is given by the formula:

$$Sociometric\ status = \frac{1}{g-1}\sum_{j=1}^{g}(\chi ji + \chi ij)$$

where g is the total number of agents in the network, i and j are individual agents, and χ_{ij} are the number of comms extant between agent i and j (Houghton et al., 2006). Sociometric status gives an indication of the prominence that each agent has within the network in terms of their ability to communicate with others. The hypothesis, therefore, is that unitary networks would generally posses fewer high-status agents (corresponding to unitary decision rights) compared to peer-to-peer networks. Specifically, the number of agents scoring more than one standard deviation above the mean sociometric status value for a given network will be higher for edge organisations by virtue of their peer-to-peer configuration than for classic C2.

Patterns of Interaction Mapped to Network Diameter

Patterns of interaction can be mapped to the social network metric 'diameter', which is given by the formula:

$$Diameter = \max_{uy} d(u, v)$$

where d(u, v) is 'the largest number of [agents] which must be traversed in order to travel from one [agent] to another when paths which backtrack, detour or loop are excluded from consideration' ($\max_{uy}$; Weisstein, 2008; Harary, 1994). Generally speaking, the bigger the diameter, the more agents there are on lines of communication. The hypothesis is that an edge organisation facilitates more direct and therefore distributed communication (and thus has a smaller diameter) than a hierarchical network, with more intermediate layers in between sender and receiver (and a higher diameter score).

Distribution of Information Mapped to Network Density

Distribution of information can be mapped to the social network metric 'density', which is given by the formula:

$$Network\ Density = \frac{2l}{n(n-1)}$$

where l represents the number of links in the social network and n is the number of agents. The value of network density ranges from 0 (no agents connected to any other agents) to 1 (every agent connected to every other agent; Kakimoto et al., 2006). It is hypothesised that an edge organisation will be denser than a hierarchical one, meaning that (all things being equal) broader dissemination of information will be rendered possible because there are more direct pathways between sender and receiver compared to a hierarchically organised counterpart.

Testing the Metrics Using Network Archetypes

The supposition that diameter, density and sociometric status can be used as metrics for decision rights, patterns of interaction and distribution of information can be tested with reference to several theoretical network archetypes. Four of these are based on early social network research by Bevelas (1948) and Leavitt (1951) who defined the following: the 'Chain', the 'Y', the 'Star' and the 'Wheel'. The value to be obtained by plotting Bevelas and Leavitt's archetypes into the approach space is consistent with the goal of identifying crucial elements of a problem space and matching regions in this space to regions in the C2 approach. Specifically, the four archetypes enable a body of empirical evidence concerning their efficacy under different task conditions to be deployed. For example, the problem space might be suggestive of a task context that is complex, with the corresponding fix within the approach space being in close proximity to the 'Star' archetype. On the basis of Bevelas and Leavitt's work, it is possible to not only make a crude judgment about this particular configuration being less than optimal but to outline more precisely why. Networks exhibiting the properties of a 'Star' often overload the heavily connected high-status node(s) in complex situations.

Bevelas and Leavitt's archetypes can be joined in the approach space by two further network structures derived explicitly from the NATO Approach Space: the 'classic C2' organisation and the 'edge organisation' (also shown in Figure 7.1). The approach space proposes that these network archetypes should in theory fall into the bottom left and top right corners respectively. The hypothesis that diameter, density and sociometric status can be used as metrics for decision rights, patterns of interaction and distribution of information can thus be subject to a direct test: if the metrics work as expected, these two network archetypes should occupy positions in the approach space predicted by the model.

Reference to Figure 7.1 shows this to be the case. The classic C2 and edge organisation, more particularly, the mapping of the relevant social network metrics to the model's fundamental axes, leads these two archetypes to indeed fall broadly into the areas of the approach space predicted. Interestingly, although classic C2 is not pushed hard into the bottom/left/front position of the space, it is positioned in the correct 'octant'. Further investigation of this phenomena reveals hierarchical networks to be much more scale dependent than comparable edge organisations, thus more realistically sized hierarchies do tend to push further into the 'correct' part of the approach space.

Broadly speaking, then, the mapping hypotheses described above is supported: diameter, density and sociometric status can be used as metrics for decision rights, patterns of interaction and distribution of information. Using these quantitative measures to plot Bevelas and Leavitt's network archetypes into the approach space also helps to fulfil the goal of identifying crucial elements of the approach/problem spaces. In conclusion, theoretical data supports the use of SNA metrics to quantify the approach space axes.

Part 2: Analysing Live NEC Using the NATO Approach Space

Live NEC Exercise

Having defined a set of social network metrics that map on to the NATO SAS-050 Approach Space, and subjected those metrics to a test with theoretical data, an opportunity now arises to scale up the analysis considerably and to test their efficacy and usefulness with realistically sized 'actual' command and control organisations. This occurs by embedding the approach developed in Part 1 above into the much bigger analysis that forms the main topic of this book. To briefly recap, the military exercise in question had the purpose of trialling a digital tactical communications and mission planning system and the current analysis formed part of a wider effort in respect of the following exploratory hypotheses:

- Exploratory Hypothesis #1: The process-like attributes of NEC, in which the human system interaction is fundamentally unstable (for example, Lee, 2001), raises the possibility of people interpreting NEC systems, amending them, massaging them and making such adjustments as they see fit and/or are able to undertake (Clegg, 2000, p. 467). The exploratory hypothesis concerns what meaningful features and/or behaviours emerge that are not anticipated by the designers of the system. The presence of such unexpected behaviours serves as a powerful indication of the type of interaction that users are trying to design for themselves and is thus a useful level of insight, especially within the context of the iterative design cycles necessary to continually evolve NEC systems towards their desired states.
- Exploratory Hypothesis #2: In the present scenario, concern had been voiced over the efficacy of the digital NEC system's interface. An analysis of comms content (carried out under the rubric of SNA) enables the content of comms to be analysed, specifically, the relationship between what Endsley (1997) refers to as 'data' (as in simple elements in the environment) versus 'information' (as in comprehension of what data means for decisions and action). The null hypothesis is that the digital interface should enable similar amounts of data to be turned into information compared to non-digital forms of comms. If an imbalance is detected (the alternative hypothesis) then the possibility that the digital system is not facilitating the data/information transformation is revealed, thus acting as a signpost for further scrutiny.
- Exploratory Hypothesis #3: Echoing the statements of Alberts & Hayes (2006) the interest is firmly directed towards where an NEC organisation 'actually' places itself. This exploratory analysis relies on deploying the social network metrics with live data to model the organisation both statically and dynamically. The aims and aspirations of the military exercise (combined with the concept of operations and standard operating procedures) is designed to render a fully net-centric force. The alternative hypothesis is that where the organisation 'actually' places itself will be different to that anticipated.
- Exploratory Hypothesis #4: The number of locations that a particular organisation is able to adopt within the approach space is a reflection of the organisation's agility, degrees of freedom or variety (Alberts & Hayes, 2006; Waldrop, 1992; Ashby, 1956). Ashby's law of requisite variety echoes one of the key objectives highlighted by the NATO SAS-050

Approach Space, which is that the area occupied by the organisation in the approach space should match that of the corresponding problem space. 'Only variety can destroy variety' (Ashby, 1956) so any mismatch is a powerful indication that the problem has the capability to overcome the approach.

Data Collection

Data collection took place at a fully functioning Brigade level Headquarters (Bde HQ) set up in the field for the purposes of evaluating a particular BP/MP system (as already discussed above). The SNA focuses on 'inter-organisational' and 'inter-cellular' comms. Inter-organisational comms took place between the Bde HQ and live, geographically dispersed Battle Group Headquarters (BG HQs). Additional data, such as further BG HQs and enemy units was simulated from an Experimental Control Centre (EXCON) which, once again, was geographically disperse from both the live Bde and BG HQs. Bde HQ in itself is a reasonably sized organisation divided up into the conceptual equivalent of 'departments' (or 'cells'). Inter-cellular comms refer to those that took place between different parts of the Bde HQ and these too placed heavy reliance on the comms capability of the NEC system. Inter-personnel comms (by non-technologically mediated means, that is, face-to-face comms) did not, and thus were not analysed on this occasion.

In all respects the Net-enabled command and control infrastructure was set up and staffed as it would have been if deployed. There were a total of 73 active agents in the scenario, 17 of whom were located in and around the Bde HQ. Agents, in this case, does not necessarily mean 'people', rather it is the number of NEC system terminals, which are more or less restricted by role but can still be used by more than one person. This number of agents and this degree of geographical dispersion creates a large-scale and complex scenario. It is important to note that the explicit aim of it was to put this particular NEC system to the test; it was not a test of the military effectiveness of the Bde and BG HQ (or any other sub-unit). As a result, the simulated enemy was probably rather more 'compliant' than that normally found, yet despite this, levels of overall complexity and tempo were very realistic.

The operations phase took place over the course of a single day (with plans and so forth being prepared the day previously) and took 4 hours and 20 minutes to complete. In broad terms it was comprised of a rapidly approaching enemy from the west who had to be steered, through a combination of turn and block effects, to the north-east of the area of operations into an area where a 'destroy' effect would be deployed. Any remaining enemy units would then continue into the next area of operations which was not under the control of the present Bde HQ.

Data Sources

Two sources of data were used to inform the analysis. Firstly, comprehensive telemetry was extracted from the NEC system. The sampling rate of the telemetry varied, but reached a maximum of approximately 10Hz and yielded a total of 2,866 data points pertaining to who was communicating to whom, as well as the broad category of 'what' was being communicated. This 'system log' data all resided at a 'digital' level in so far as it presented itself to the user through the NEC system's data terminals.

The second source of data was voice comms which were transmitted over the encrypted radio part of the NEC system. Data collection relied on a formal log of those comms kept by the incumbent of the Watch Keeper role. Every communication, its time, from who it derived and to whom it was directed, and its content, was recorded and this formed the basis of an analysis of inter-organisational 'voice' comms. Although mediated by a digital radio technology the presenting modality of the communication, from the users point of view, was 'voice'. A total of 158 discreet events of this type were extracted.

Modelling

The broad remit, scope and nature of SNA has been described above in general terms. The time has now come to describe in rather more explicit terms exactly how the data was treated and how the analysis was undertaken, including a number of important extensions.

Digital and Voice Layers: The first step was to keep the 'digital' and 'voice' comms data separate and to define them as two distinct functional architectures. The reason for this is that the contrast between the different layers is an important part of the analysis, after all, different means of communication are likely to convey different content, occur at different times, at different frequencies and have different issues.

Communications Content: The next step in the analysis was to perform a high-level analysis of comms content. The frequency of transmit and receive events was considered first, followed by a slightly more detailed breakdown of comms content. At the digital comms layer the system logs already provided a comms 'type' label and this was preserved and subsequently interpreted in the analysis. The voice comms layer required a different encoding taxonomy and a categorisation more appropriate to this layer was derived from Bowers et al. (1998). Seven commuications typologies are defined, with every instance of an inter-organisational verbal exchange being categorised as either:

- Factual: 'objective statement involving verbalised readily observable realities of the environment, representing "ground truth"'.
- Meta-query: 'request to repeat or confirm previous communication'.
- Response: 'statement conveying more than one bit of information' (that is, comprising of more than simply 'yes/no').
- Query: 'direct or indirect task-related question'.
- Action: 'statement requiring team member to perform a specific action'.
- Acknowledgement: 'one bit statement following another statement (for example, 'yes', 'no')'.
- Judgment: 'sharing of information based on subjective interpretation of the situation' (Cuevas et al., 2006, p. 3–4).

'Data' versus 'Information': Clearly, with two distinct categorisation schemes extant across digital and voice comms layers, a way to compare across them is required. For this a cautious distinction between 'data' and 'information' is made. Caution is exercised because defining terms like these represents something of a popular game for Ergonomists, and theoretically at least, these terms remain contentious. As a result, nothing more than a crude attempt to impose this categorisation is made currently, and even then it is intended that 'data' and 'information' serve merely as broad descriptive labels. A tolerable stab at defining these terms is made by anchoring them to Endsley's three stage model of SA (1988).

Endsley (1997) points out that success in something like the Bde HQ's endeavours 'involves far more than having a lot of data. It requires that the data be transformed into the required information in a timely manner', furthermore, 'creating information from data is complicated by the fact that, like beauty, what is truly "information" is largely in the eyes of the beholder' (p. 2). In defining these terms we turn to Endsley's model of Situational Awareness (SA). Here, Level 1 SA refers to 'the perception of elements in the environment' and for current purposes emphasis is placed on 'elements in the environment'. These we might refer to simply as 'data', that is, as objective, measurable realities of a situation. Level 2 SA refers to comprehension of what those elements might mean (thus, in the eye of the beholder, mere data is becoming information) whilst Level 3 SA refers to 'projecting their status into the near future' (in other words, actually doing something with the information). Grouped together, Level 2 and 3 SA seem to bound a category that is undoubtedly more 'information-like' than it is 'data-like' (or Level 1 SA-like).

Static Network Modelling: The raw comms data (both digital and voice) was split into blocks of 80 comms events. These blocks of data were then converted into social network diagrams using a proprietary software tool called WESTT (Workload, Error, Situation Awareness, Tasks and Time; Houghton et al., 2006 & 2008). This resulted in 34 separate social networks being created for the digital comms layer and 30 for the voice comms layer. Associated with each network was, of course, an accompanying density, diameter and sociometric status metric. The social network metric data was split in half into 50th percentiles to create a 2 x 2 x 2 taxonomy of: diameter = distributed versus hierarchical, density = broad/versus tight, and status = unitary versus peer-to-peer (see Table 7.1 for an example). The 50th percentile containing the greatest number of data blocks represented the modal or static social network achieved in the digital and voice layers. In other words, it represents the organisation's centre of gravity, a static time independent characterisation of the organisation's fundamental mode of operation. The centre of gravity for the various network archetypes was also computed using the WESTT software tool in order to facilitate a comparison between live data and theoretical data.

Dynamic Network Modelling: The organisation's movement about its centre of gravity was visualised by using each of the 30 (voice) or 34 (digital) social networks to plot the organisation into the approach space. By these means it was possible to see how its location varied across time (with each data block representing a time interval) and function (voice versus digital layers). The range of values that density, diameter and high-status nodes describe over time say something about the organisation's agility. When these same values are each plotted on to a graph with the x-axis representing time and the y-axis representing the corresponding social network metric value, the resultant line resembles a waveform. Spectral analysis methods can then be deployed in order to say something about 'tempo'.

Spectral analysis methods, broadly speaking, decompose the data into frequency components using Fourier Analysis. The principle here is that any complex waveform can be separated into its component sine (or pure) waves and the resultant data plotted into a graph called a Periodogram. A periodogram has a frequency scale on the x-axis measured in Hertz (or cycles per second) so that moving right along the x-axis the higher the frequency, and more rapidly fluctuating the density, diameter or status value becomes. But as well as speed of fluctuation there is also the strength of that fluctuation, which is ascribed a power value on the y-axis. The higher the power value is, the stronger the fluctuation at that particular frequency. Such an analysis, therefore, provides a surprisingly literal representation of 'battle rhythm', albeit subject to a modifying caveat. The measurement intervals used in the current analysis distort the frequency scale somewhat (the sampling rate is not in cycles per second). This means that it is not safe to interpret the x-axis of the periodogram literally but rather to interpret it 'relatively'.

With this caveat in place, the periodogram represents an excellent diagnostic tool in terms of it being able to detect the presence of meaningful temporal patterns in the way that the organisation configures and reconfigures itself under real-life dynamic conditions. If meaningful patterns exist then something of their character can be revealed: evidence of periodicity indicates that the organisation is repeatedly drawn into specific locations of the approach space (as a fixed attractor in phase space draws physical systems into defined equilibrium states). However, the lack of a pattern or of a periodicity is just as interesting. It suggests that the forces present in the problem space which are driving the organisation's dynamic reconfigurations are more chaotic in nature (referred to as 'strange attractors' in complex systems terms).

Noise Reduction: A final point is made by Dekker (2001; 2002) that SNA, as is common in the social sciences, often contains a significant element of random noise. Steps have to be taken to provide a measure of reassurance that the effects being observed are robust rather than an attribute of random error. To this end the spectral analysis technique just mentioned is combined elsewhere in the analysis with various inferential statistics, the aim in both cases is to go beyond mere description and to try and draw out 'meaningful' patterns in the data.

To sum up the approach to analysing the live data: firstly, the raw comms data is split into digital and voice layers. Secondly, each layer (digital and voice) was subject to its own form of comms content analysis. Thirdly, the comparison between digital and voice layers was facilitated by a generic data versus information categorisation. Fourth, the organisational centre of gravity for both digital and voice layers was derived followed by, fifth, splitting the data into blocks in order to plot the dynamic behaviour of the organisation into the approach space in order to reveal its agility. Sixth, and finally, the dynamic data was deployed further in order to perform a spectral analysis of tempo. This six step process of data analysis enables all of the exploratory hypotheses to be explored.

Results: Content of Communications

Digital Communications

2,866 individual digital comms events took place over the course of the 4 hour 20 minute exercise. This equates to a mean of 662 comms per hour (or approximately 11 per minute). Of those 2,866 events, 454 were transmissions (that is, data leaving the Bde HQ) and 2,412 were receive events (that is, data entering the Bde HQ). In order to statistically check that this marked disparity is a non-random effect a Binomial test is performed. In percentage terms only 16 per cent of digital comms were transmissions compared to 84 per cent of receive events. Statistical assessment of this disparity allows us to accept the hypothesis that this is a non-random effect ($p > 0.0001$, two-tailed): digital comms are predominantly 'received' rather than transmitted.

The pie charts in Figure 7.2 break down the type of comms further in order to examine its content in more detail. The content of what is being transmitted is based on the descriptive label provided by the System Logs that generated the data.

The largest proportion of transmissions is accounted for by System Messages. These relate to the system's status and current activity (37 per cent). This is followed by acknowledgements that something has been received, or of an action that has been performed on the system (30 per cent). Next in line, and perhaps most interesting in the present context, is Free Text (9 per cent). Free Text, as the name implies, is the facility for users to type messages without the imposition of any kind of pre-determined layout, format or proforma. In other words, it is a catch-all form of functionality which, given the level of structured interaction provided by the system, is not one that the designers seemed to anticipate being used all that extensively. However, in practice, it seems clear that personnel rely on it to a far greater degree than expected. Further Binomial tests demonstrate a persuasive non-random effect when Free Text is compared to various other forms of 'constrained text' (that is, comms in the digital domain that are constrained by templates, pre-defined formats and so on). Table 7.1 below presents the results of this comparison.

The pie charts in Figure 7.2 show that the proportion of Free Text comms is statistically greater than any other individual form of constrained communication. So, for an 'additional

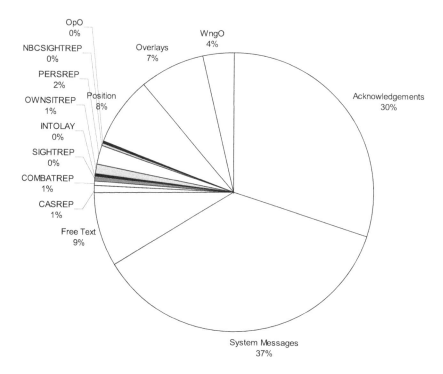

Figure 7.2   **Pie chart showing the 'type' of data communications being transmitted (note that 0 per cent indicates a percentage of less than 1 and that narrow segments have been shaded for legibility)**

Table 7.1   **Free versus 'Constrained' Digital Communications that are 'Transmitted'**

	N	Observed Proportions	
Free Text	39	Constrained	FreeText
CASREP	4	0.09	0.91
COMBATREP	3	0.07	0.93
SIGHTREP	2	0.05	0.95
INTOLAY	2	0.05	0.95
OWNSITREP	4	0.09	0.91
PERSREP	11	0.22	0.78
NBCSIGHTREP	1	0.03	0.98
OpO	1	0.03	0.98

feature', one whose use is not actively encouraged in standard operating procedures, it seems to have assumed a much more important role than anticipated. To that extent, at least, it can be considered emergent.

Moving from comms transmitted to those that were received (n=2412), Figure 7.3 breaks down these events into their content where it is clear that they differ considerably in character from those that are transmitted. By far the largest proportion of incoming digital comms is accounted for by positional data broadcast by sub-units (58 per cent). Note that the system uses this data to display icons on a map. This is followed by System Messages (25 per cent) and Acknowledgements (13 per cent). It is interesting to note again the relative prominence of 'Free Text'. It is fourth highest proportionately, albeit comprising only 3 per cent of total comms. Nevertheless, a similar finding to the above is detected in that Free Text accounts for a greater proportion of comms than do the more constrained forms of text comms as Table 7.2 shows.

In summary, the findings suggest that the quantity and content of digital comms varies considerably with its direction. Broadly speaking, by looking at the two biggest categories, positional data is received whilst acknowledgements are sent/transmitted. Perhaps the most interesting finding revealed by the present analysis is the reliance placed on the Free Text facility. Exploratory hypothesis #1 is supported in that this effect seems to have emerged as a product of user adaptation and is somewhat contrary to how the system was expected to be used. Despite this, it serves as an indication of the kind of interaction that users were trying to design for themselves, and evidently, they prefer a much less constrained form than is currently on offer.

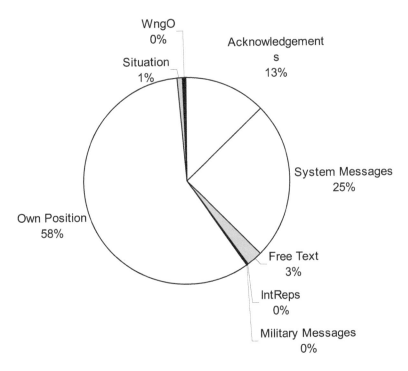

Figure 7.3 **Pie chart showing the type of data communications received (note that 0 per cent indicates a percentage of less than 1 and that narrow segments have been shaded for legibility)**

Table 7.2   Free versus 'Constrained' data communications that are 'Received'

	N	Observed Proportions	
		Constrained	FreeText
Free Text	65		
INTREPS	4	0.03	0.97
Military Messages	3	0.06	0.94
WngO	2	0.16	0.84

Voice Communications

Data for the voice comms analysis is derived from the Watch Keeper's log, a formal record of incoming and outgoing radio comms. A total of 158 comms events occurred over the exercise period which equates to a mean of 36 voice comms instances per hour (or approximately one every 2 minutes). As is the case for digital comms, there is a marked disparity in the number of transmissions (n=45) versus receive events (n=113). Applying the same Binomial test procedure leads us to accept the hypothesis that this disparity did not arise as a product of random error alone ($p < 0.0001$, two-tailed).

The same comms 'content' categorisation does not apply across digital and voice layers and as mentioned above, a categorisation for the voice comms data is derived from Bowers et al. (1998). The results of this categorisation are shown for transmit events and receive events (see Figures 7.4 and 7.5).

The sum total of these categories represents a 'total communications' score and the pattern of comms can be characterised statistically with reference to a multiple regression-based technique (Warm et al., 1996). By this method, each individual category is analysed in terms of its relative contribution to the aggregate 'total communication' score. Its contribution is expressed in terms of a standardised beta coefficient, the higher the beta coefficient, the more that a comms subscale contributes to total comms. This method provides an order of magnitude analysis highly useful for discerning patterns of communication such as this. Note also that an assessment of statistical significance is carried out on the beta-coefficients thus providing a means to assess the affect of random error and noise in the data (a statistically significant coefficient value is one that is unlikely to have arisen due to random error). The pattern of relative contributions made by each subscale varies across 'receive' and 'transmit' events, as shown in Table 7.3.

What does the table above communicate about the pattern of voice comms? First of all is the presence of a marked change in pattern between transmit and receive. Meta-Queries, Responses, Queries, Actions, Acknowledgements and Judgments make a uniform contribution to overall comms in the transmit category, with Factual comms assuming a lesser place. In other words, the output of Bde, in voice comms terms, is less about 'objective statements involving verbalised readily observable realities of the environment' and rather more to do with 'requests to repeat or confirm previous communication, statements conveying more than one bit of information (that is, comprising of more than simply 'yes/no'), direct or indirect task-related questions, statements requiring team members to perform a specific action, one bit statements following another statement (for example, 'yes', 'no') and the sharing of information based on subjective interpretations of the situation'. This pattern switches for comms that are received.

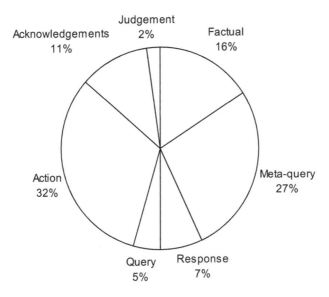

Figure 7.4   Pie chart showing the content of voice communications transmitted according to Bowers et al.'s (1998) taxonomy

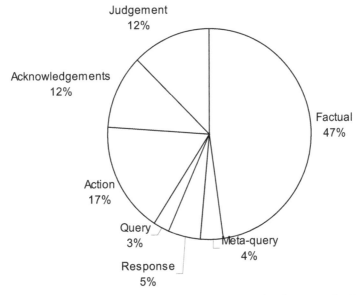

Figure 7.5   Pie chart showing the type of voice communications received according to Bowers et al.'s (1998) taxonomy

Here, the focus is very much on observable realities of the situation. The picture that emerges from this analysis is one of a 'data' versus 'information' dichotomy, with the former associated with comms received and the latter associated with comms sent (see Table 7.4).

Table 7.3   Standardised beta coefficients show the relative contribution that each voice communications subscale makes towards total communications (for both transmit and receive events)

	Beta Coefficients		
	Receive	Transmit	Direction of Change
Factual	0.73*	0.33*	Down
Meta-Query	0.17**	0.67*	Up
Response	-0.30	0.67*	Up
Query	0.16***	0.67*	Up
Action	0.60*	0.67*	Up
Acknowledgement	0.40*	0.67*	Up
Judgement	0.43*	0.67*	Up

* Statistically significant at the 1 per cent level

** Statistically significant at the 5 per cent level

*** Statistically significant at the 10 per cent level

'Data' versus 'Information' (sic)

The analyses shows all comms (digital and voice, both transmitted and received) divided into 'data' and 'information' meta-categories. As defined in the Methodology section above, data is taken to mean an emphasis on raw 'elements in the environment' (Level 1 SA according to Endsley's model) whereas information emphasises comprehension and projection or the simple expedient of 'something having happened to it in the mind of an individual' (Level 2 and 3 SA). This shows what the different comms layers (digital and voice) predominantly carry in terms of this. From the table it is clear that the digital comms layer carries a far higher proportion of data compared to the voice layer, which carries a greater proportion of information. This disparity is statistically significant ($\chi^2 = 14.08$; df = 1; $p < 0.0001$, and is accompanied by a moderate effect size; Cramer's V = 0.27; $p < 0.0001$). Note that the Chi Square test has been performed on the percentage of the raw scores. The reason for this is that statistical power becomes excessively high with 3,000 or so data points thus making it extremely unlikely that a statistical difference 'wouldn't' be detected. Table 7.5 contributes further to this picture. It goes on to show that it is predominantly 'data' that is received by the Bde HQ (96 per cent) compared to a greater proportion of 'information' leaving it (26 per cent).

The pattern of findings was once again subject to statistical tests to assess the probability that they did not arise due to random error. In this case $\chi^2 = 18.98$; df = 1; $p < 0.0001$ which means that there is a significant association between the direction of comms (transmit or receive) and their type (data or information). Or in other words, the observed results differ significantly and

Table 7.4   Data and information carried by digital and voice communications

	Digital %	Voice %
Level 1 ('data')	73	47
Levels 2 & 3 ('information')	27	53

Table 7.5   Data vs Information Received and Transmitted by Brigade Headquarters

	Received by BDE	Transmitted by BDE
Level 1 ('data')	96	74
Levels 2 & 3 ('information')	4	26

meaningfully from the results expected under the null hypothesis (for example, Cramers $V = 0.31$; $p < 0.0001$, a medium effect size).

Overall, what these findings mean is that the incoming data arrives on the digital layer as data and leaves the Bde HQ on the voice layer as 'information' with the system's interface sitting squarely in between. From the imbalance between data/information, the interface appears to act as something of a bottleneck. The alternative hypothesis (of exploratory hypothesis #2) is supported. A possible design goal of future versions of the system might, therefore, be to increase the proportion of 'information' on the digital layer from a lowly 27 per cent to something more like the voice layer (approximately 50 per cent). To express this in more colloquial terms, this is a lot of data (and a lot of bandwidth) for not much resultant information.

In wider social network terms it is interesting to note that the behaviour of the network, specifically the pattern of receive and transmit events, certainly does resemble 'a model of a biological organism' (Dekker, 2002, p. 95) in that 'the organisation receives [data] from its environment (intelligence), makes decisions, and produces some effect on its environment (force). [...] Ultimately, the performance of an organisation (or an organism) depends on the appropriateness of its response to its environment' (p. 95). The corollary of this is that the environment is dynamic, as should the corresponding behaviour of the social networks be. This biological conception of what a social network represents necessarily embodies the language and metaphors of systems theory, specifically 'open systems theory', and is thus a highly appropriate conceptual response to an organisation like NEC that embodies open systems principles like agility and self-synchronisation. It is to this type of analysis that the next section of the results now turns.

Results: Structure of Communications

Digital Communications

Static characterisation   The data from the scenario that shows who is communicating with whom was broken down into 34 individual stages, with each stage being subject to a SNA. This involved constructing a matrix (identifying who was communicating to whom and how often), producing network diagrams, and from them, computing various network statistics which in turn map on to the NATO Approach Space's three main axes.

In order to perform a static characterisation of the organisation's centre of gravity within the approach space it is divided into eight 'octants'. By this reckoning social networks can either be broad or tight (in terms of decision rights), distributed or hierarchical (in terms of patterns of interaction), unitary or peer-to-peer (in terms of dissemination of information), or any combination thereof. It is worth pointing out that it is considerably more difficult to make a similar diagnosis when inspecting network diagrams in their raw form, especially those embodying the current level of complexity.

The 34 separate social network analyses performed on the digital comms layer produced 34 separate diameter, density and sociometric status figures. These were then divided into upper and lower 50th percentiles to create six categories, as shown in Table 7.6. The raw data was then transformed into category data (1 = Upper Percentile, 2 = Lower Percentile). The number of data points that fell into each category was then subject to a simple modal analysis to derive an 'average' network type. It will be noted that none of the categories are strongly biased in any particular direction, so this high-level characterisation is undeniably of a broad brush nature. The results of this analysis are shown in Table 7.6.

The modal network derived from the above analysis was able to be compared against the earlier network archetypes, the 'Circle', the 'Chain', the 'Y' and the 'Wheel'. The results of this comparison are also shown in Table 7.6.

Bevelas (1948) and Leavitt's (1951) prototypical networks (and the performance characteristics associated with each) can now be used to meet one of the central objectives of SNA; to draw out the implications of the digital comms network observed at Bde HQ. It can be seen that the static characterisation of the Bde digital comms network approximates most closely to the 'Circle' network archetype, which is to say that the centre of gravity for this layer of comms is located in that vicinity. Bevelas and Leavitt's work highlights the performance advantages of this configuration under situations of high task complexity. In these situations the decentralised nature

Table 7.6   **Overall characterisation of the network type extant at the digital communications layer compared to a number of social network archetypes**

		Diameter	Density
Digital Layer	Upper Percentile	16	17
	Lower Percentile	18	17
	Modal Point	Hierarchical	Broad/Tight (Tied)
Archetypes	Peer-to-Peer	Distributed	Broad
	Hierarchical	Distributed	Tight
	Circle	Hierarchical	Broad
	Chain	Hierarchical	Tight
	Y	Hierarchical	Tight
	Wheel	Distributed	Broad

* Shading denotes closest match

of the network helps to avoid bottlenecks and the overloading of just one, or of a few, heavily connected agents. In terms of exploratory hypothesis #3 this is undoubtedly where the organisation 'actually' places itself, and in this context is isomorphic with where the organisation desires to place itself (as a decentralised net-centric organisation). However, consistent with exploratory hypothesis #4, the mapping between its location in the approach space and the corresponding location in the problem space reveals a mismatch (thus the alternative hypothesis #4 is supported). Plotting the archetypical networks directly into the NATO approach space as shown in Figure 7.6 locates the 'Circle' archetype in the bottom/right/back octant. The problem, that is, the scenario, is an overtly cold-war style of engagement characterised by high familiarity (of enemy doctrine), fairly static rates of change (to the extent that the dynamics are more or less linear and in sequence) and a high-strength information position (a lot is known about enemy capability and position). The corresponding octant in the problem space is, therefore, in the region of the bottom/left/front octant. In this situation, the attributes of a circle network are less optimal than hierarchies, chain and Y networks, because as the resultant problem complexity decreases the time taken to collate information begins to negate the benefits of decentralisation. This precise phenomenon was clearly in evidence during the exercise and partly one of the reasons why users reverted to Free Text (as in hypothesis #1) in an attempt to speed things up.

Dynamic characterisation If the comms network was stable then the high-level static characterisation described above would be sufficient. Unfortunately, as the category data suggested (by not revealing a particularly strong category bias) the network is far from stable, in fact, it reconfigures dramatically in response to its environment. This reconfiguration is clearly evident when all 34 sequential social networks from the observed exercise are plotted into the approach space along with the archetypal networks. What results is a form of 3D scatter plot or 'phase space' that illustrates the dynamical behaviour of the network over time. What is immediately clear, and perhaps contrary to the impression often given, is that the command and control organisation, in social network terms, is highly dynamic. It does not occupy a static position within the approach space, rather it moves around it in response to the environment. Although the circle network archetype represented something of the desired or doctrinal definition of the organisation, the 'actual' region in the approach space adopted by the organisation proved to be quite different over time. The extent of the organisation's agility can be seen in that the digital comms network density varies about its mean of 8.36 by +/- 4.96. Similarly, the number of high-status agents in the network varies from zero to four, and likewise, diameter varies around 4.38 by +/- 3.5. These are pronounced changes in the structure of the social network as illustrated in Figure 7.6.

A periodogram of network density is shown in Figure 7.7. This is the output of the spectral analysis method described earlier and is designed to detect the presence and strength of any periodicity in the data. The x-axis (marked frequency) represents a scale with very slow fluctuations in digital comms density at the leftmost end, to rather more rapid fluctuations at the other. The pattern of results gained is intriguing. They show a power spectrum dominated by very low speed fluctuations in network density accompanied by the presence of what appear to be second and third order harmonics of diminishing strength but increasing frequency. In other words, overlain on this slowly cycling level of network density is a higher frequency of network reconfiguration but at reduced power. Overlain on top of this is yet another higher speed reconfiguration. In the language of signal processing the time series data seems to exhibit a fundamental frequency of reconfiguration (a strong periodicity) combined with second and third order harmonics at multiples of that fundamental. What does this mean? It means, simply, that network density exhibits a strong, non-random and often rapid pattern of reconfiguration, with comms links able to establish (and re-establish) themselves repeatedly. This seems to indicate that the digital comms network possesses

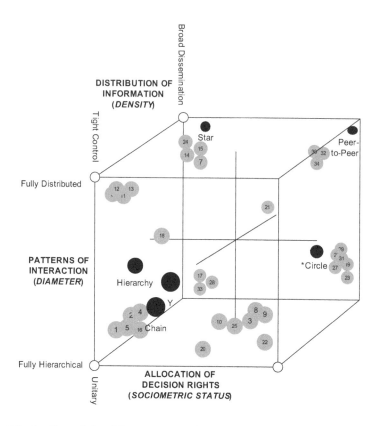

Figure 7.6   **Illustration of the 34 separate social network analyses plotted into the NATO SAS-050 Approach Space to show how the configuration of digitally mediated communications changes over time (grey numbered spots). The approximate position occupied by the network archetypes is also shown (black annotated spots). The 'modal' network configuration of the digital communications layer approximated to a 'circle' archetype (marked with an asterix)**

high tempo on this dimension. More fundamentally, the underlying theory governing this dynamical behaviour appears to be rather more deterministic in character.

Network diameter also exhibits interesting behaviour (see Figure 7.8). Again, there is the presence of what appears to be a strong fundamental frequency (suggesting an underlying rate of expansion and contraction in the network's diameter) with something that could be taken for a second harmonic occurring at a multiple of the fundamental higher up the frequency range. What this graph communicates is that there are strong and non-random changes in the patterns of interaction, with comparatively rapid shifts occurring between fully hierarchical and fully distributed modes of operation. Again, high tempo seems in evidence.

The pattern of results observed above for network density and diameter is not similarly evident for sociometric status. The power spectrum (not reproduced in this case) exhibits a more or less random pattern, a lower, more uniform power spectrum, and little evidence of systematic periodicity. In these terms, there is evidence that movement along the scale from unitary to peer-to-peer decision rights is rather low in tempo (relatively speaking) but governed by more non-deterministic factors that lack distinct patterns and fixed attractors in the phase/approach space.

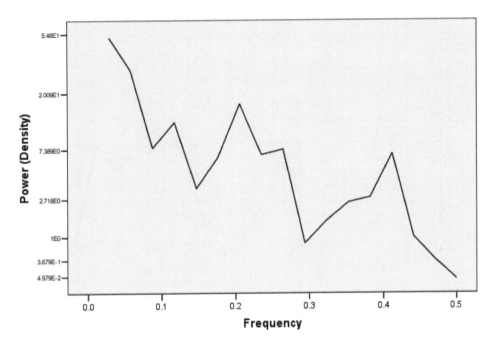

Figure 7.7 **Periodogram illustrating the presence of periodic changes in network density. A pattern is obtained that approximates to first, second and third order harmonic effects**

Voice Communications

The analysis performed on the digital comms data can now be repeated for voice comms. As a lot of the explanatory ground work has already been covered above, this section can be considerably briefer and to the point.

Static characterisation   The first stage of the analysis is to provide a static representation of the underlying voice comms data by undertaking a simple form of modal analysis, as before. The results of this are shown in Table 7.7.

It can be seen that the static characterisation of the Bde HQ voice comms network, like the digital network, approximates most closely to the 'Circle' archetype. The advantages of this configuration under situations of high task complexity have already been noted and apply here. But again, this appears to be a relatively poor match to the extant situation, perhaps to a slightly greater extent even than the digital comms layer. In simple terms, for the scenario being faced by the organisation it probably needs to be locating itself in the region of the 'hierarchy', 'chain' and 'y' archetypes which in practice it does not, as shown in Figure 7.9.

Dynamic characterisation   The reconfiguration of the voice communiations network over time is clearly evident when all 30 sequential social networks are plotted into the NATO Approach Space. In the case of the voice comms network, the density varies about its mean of 2.57 by +/- 9. Similarly, the number of high-status agents in the network varies from zero to two, and likewise, diameter varies around 1.06 by +/- 2. These changes, when summed, seem to bound a

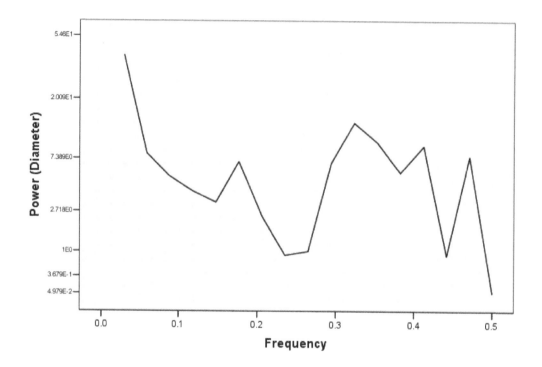

Figure 7.8   **Spectral analysis graph illustrating the presence of periodic changes in network diameter**

Table 7.7   **Overall characterisation of the network type extant at the voice communications layer compared to hierarchical and peer-to-peer archetypes**

		Diameter	Density	Status
Voice Layer	Upper Percentile	18	17	18
	Lower Percentile	16	17	16
	Modal Point	Hierarchical	Broad/Tight (Tied)	Peer to Peer
Archetypes	Peer-to-Peer	Distributed	Broad	
	Hierarchical	Distributed	Tight	
	Circle	Hierarchical	Broad	
	Chain	Hierarchical	Tight	
	Y	Hierarchical	Tight	
	Wheel	Distributed	Broad	

* Shading denotes closest match

similar 'sized' area in the approach space as the digital comms layer, so to that extent both layers display similar outright levels of agility. It is important, however, to note that the shape of the area plotted within the space does appear to be different. Agility, therefore, can be the same (in area) but different (in shape) as is the case here. The visual representation provided by the NATO Approach Space reveals some visual differences to support this.

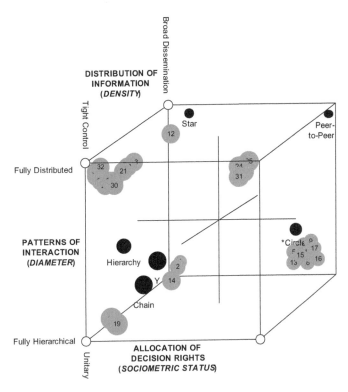

Figure 7.9 **Illustration of the 34 separate social network analyses plotted into the NATO SAS-050 Approach Space to show how the configuration of voice mediated communications changes over time (grey numbered spots). The approximate position occupied by the network archetypes is also shown (black annotated spots). The 'modal' network configuration of the voice communications layer approximated to a 'circle' archetype (marked with an asterix)**

Reference to Figures 7.6 and 7.9 shows that the voice layer appears 'tighter' overall, the individual networks tending to fall into more well defined areas of the 3D space compared to digital comms, which is looser, with networks dispersed more widely on some axes. One way to interpret this is to say that the voice comms phase space is characterised by powerful fixed attractors that control the dynamics of the organisation in more prescribed ways. The phase space described by the digital comms layer lacks these fixed attractors, characterised instead by strange attractors and more chaotic underlying dynamics. Spectral analysis may help to shed more light on this.

Figure 7.10 is a periodogram of network density for the voice comms layer. The pattern of results is more or less identical to those gained for network diameter (given the similarity and the order of magnitude interpretation of the periodogram, only that for density need be reproduced). This shows a power spectrum dominated by a very low speed fluctuation in network density and diameter, one that rather corroborates the 'larger steps' these networks make in terms of their reconfiguration and indicative of relatively low tempo on this dimension. It communicates the fact that the network configures and reconfigures itself, and is able to alternate between 'hierarchical' and 'fully distributed', and from 'tight' to 'broad dissemination' modes. However, compared to the digital comms layer, there is a relative absence of high frequency harmonics

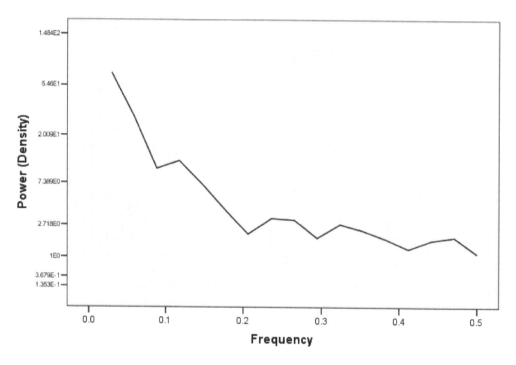

Figure 7.10   **Spectral analysis graph illustrating the presence of periodic changes in network density. A more or less identical pattern of results is achieved for network diameter**

(or other components) and it is further interesting to note that the power level is substantially reduced. So, while there is evidence of a non-random periodicity in diameter and density, the tempo of the voice comms network is neither as rapidly configuring nor as powerful as that for the digital comms layer. This can be seen as a success for agility and tempo embodied by the net-enabling technology.

Unlike the digital comms layer, the behaviour of the network in terms of high-status nodes is interesting and of an order of magnitude greater in terms of power (see Figure 7.11). The pattern of results is difficult to determine but unlike the previous case there are two discernable peaks in the spectrum that indicate a low and high speed change in the number of high-status nodes. On this measure, then, the voice comms network appears to be governed by more deterministic forces than the comparable digital layer (which exhibited greater degrees of chaotic behaviour).

To summarise, agility and tempo differ according to the functional division between digital and voice comms. The behaviour of the digital system seems to exhibit greater agility in that the number of distinct points in the phase space that it can adopt are greater and more widely dispersed than those present on the voice layer, despite both layers being comparable in terms of the total area occupied within the phase space. Generally speaking, there are also several key differences in the nature of the organisational dynamics, with behaviours along the three primary NATO Approach Space axes governed by different degrees of determinism/chaos depending on function (digital versus voice). In both cases, however, the actual locations adopted by this live instance of NEC within the approach space struggled to match the appropriate locations in the

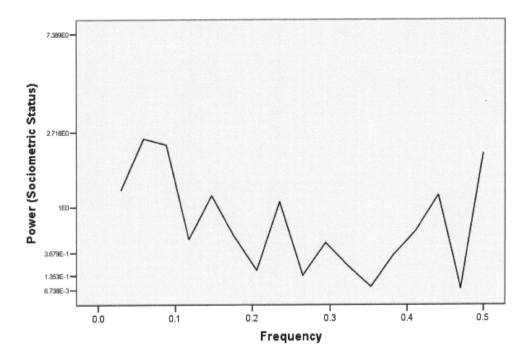

Figure 7.11 **Spectral analysis graph illustrating the presence of periodic changes in high-status nodes**

corresponding problem space. One side effect of this seems to have been the emergent use of Free Text as a way to get the system behaving the way users desired.

Summary

The key innovation presented in this chapter is to use SNA in order to define quantitative metrics for each of the NATO Approach Space's primary axes. It is this simple expedient that has provided a practical means to define the actual place or region in the approach space where an organisation operates, to see how that ranges over function (digital and voice layers) and time (by taking numerous slices through the data). In addition, through the use of well understood network archetypes, a comparison between the approach and the problem to which it is directed has been facilitated. In summary, the use of both theoretical and live data demonstrates how the missing links of the NATO Approach Space can be completed.

Of course, there is still work to do on what is admittedly a relatively fresh innovation. Future work is directed into the following areas: further refinement of the mapping between the NATO Approach Space's axes and social network metrics, exploration of issues such as how these metrics behave at different scales, how concepts from complex systems research can help to understand and model the underlying dynamics of NEC systems, and how a similar quantification approach can be performed on the 'problem' space. These and other issues are all being vigorously pursued. For the time being, though, this work is also offered up as a benchmark for researchers, analysts and experimenters engaging in C2-related research.

Chapter 8
SCADA Analysis

EEMUA 201 SCADA System Analysis

The following analysis is based upon EEMUA 201:2002. This guide from the Engineering Equipment & Materials Users Association (EEMUA) provides a checklist for designing and evaluating Supervisory Control and Data Acquisition (SCADA) systems. The guide is primarily aimed at more constrained civilian process control systems such as gas and electricity distribution; however, the activities of Battle Group (BG) and Brigade (Bde) also fit into this definition. The role of the Headquarters (HQ) is to supervise and control subordinate units. This task is assisted through data acquisition and dissemination.

The data was collected in a team of three; two Human Factors professionals (authors of this book) and a Subject Matter Expert (SME). The SME was an experienced trainer of the digital Mission Planning and Battlespace Management (MP/BM) system with an expert understanding of the tool along with its functionality and capabilities. The data collection exercise lasted approximately 90 minutes. The requirements were approached in the order presented, the SME controlled the terminal and the screen of the terminal was duplicated on a projected screen. Where possible evidence was produced relevant to the EEMUA requirements, the evidence was recorded by the Human Factors professionals textually and accompanied by screenshots of the process. The SME agreed to each of the statements made on compatibility with the EEMUA guidelines.

The number of screens must allow for complete access to all necessary information under all operational circumstances

Figure 8.1 shows the main window of the system which displays the Local Operatinal Picture (LOP). The screenshot presented in Figure 8.2 illustrates that when an Operational Plan window is opened, the LOP is obscured. The current system, with its single display does not support the requirement of monitoring the LOP whilst conducting staff work. Specific staff work activities are actor dependent; they include activities such as: product creation; message sending and receipt; and data handling.

To compensate for the single display some users (for example, the Ops Officer at the BG main HQ) have two screens available to them; however, this is through the use of two terminals. Although the terminals are networked and thus allow some level of data transfer they are essentially separate systems. The user must use two separate terminals and thus two separate input devices; the mice and keyboards of two terminals.

The Design should Allow for a Permanently Viewable Overview Display Format

Figure 8.1 shows the overview display format, but as highlighted in the previous requirement this screen is frequently obstructed when the user needs to carry out any staff work. Wherever possible, two terminals should be used in order to allow for a permanently viewable overview display format. Figure 8.3 illustrates the two screen system, with one screen displaying the LOP and the other displaying a planning system.

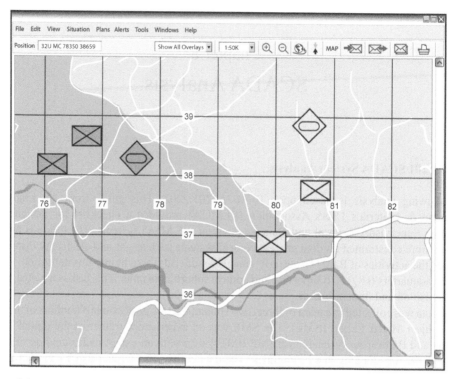

Figure 8.1   Diagram showing the main system screen (Local Operational Picture; LOP)

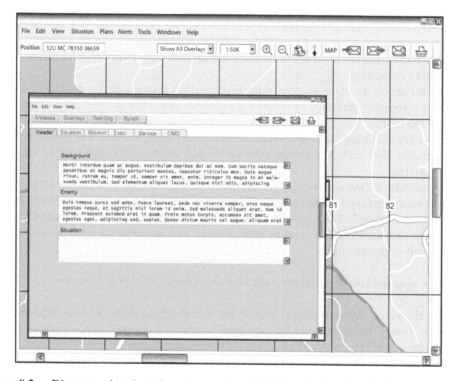

Figure 8.2   Diagram showing the obscuration of LOP window by the Operational Plan window

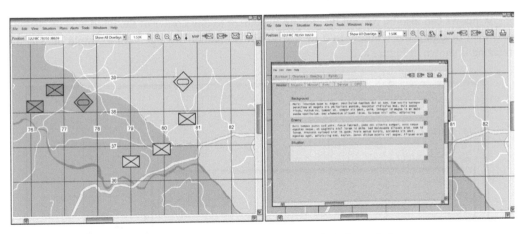

Figure 8.3   Diagram showing the LOP window and a planning window

Continuous Access to Alarm Information should be Provided

There is no dedicated alarm screen; pertinent messages can be transmitted as 'flash' messages. These always appear in the centre of the screen flashing.

The Capability to Expand the Number of Screens should be Built into the Design

The capability to expand the number of screens is available in the form of adding more terminals. Figure 8.3 shows the two windows needed by users. There are a number of usability concerns with this system as a result of the duplication of input devises. Users are likely to confuse the input devices frequently. Users may also suffer from physical discomfort through the positioning of these devices. The user will have to make frequent posture alterations to successfully use both terminals.

It is also postulated that a system made up of two terminals is financially less cost effective than a terminal with two screens.

The Provision of Screens should Allow for Start-up, Steady State Monitoring, Shut-down and Abnormal Situations

It is possible to have separate screens for these four states; however it would involve the use of several separate terminals networked together. The system itself does not allow for multiple screens.

The System should Allow the User to Resize Windows Easily

With respect to the resizing of windows, this system has functionality similar to 'Microsoft Windows'. It is possible to resize windows using the minimise and maximise buttons; however, this functionality may invoke a problem for terminals where the user has only a stylus as these resize buttons are located directly next to the close window button. Thus it would be very easy for the user to accidentally close the window instead of resizing it.

Windows can also be resized by moving the cursor to any corner of the window, left clicking and then dragging the window to a new size, either smaller or larger. An example of window resizing is shown in Figure 8.4.

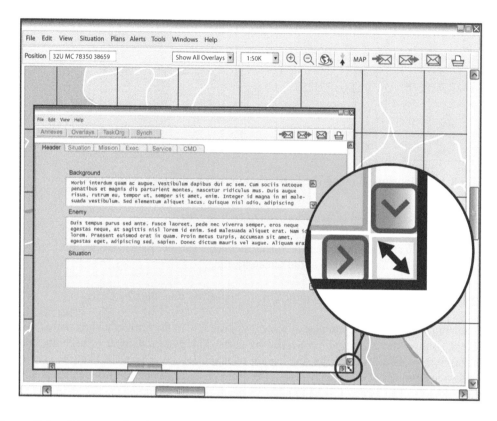

Figure 8.4   Diagram showing the ability to resize windows

The System should Enable Ease of Navigation Up and Down the Hierarchy of Screens

The system refresh rate is too slow to allow for easy navigation up and down the hierarchy of screens. The system refresh rate should be fast enough to enable ease of navigation.

The System should have a Simple Hierarchical Structure, with a Total Plant Overview at the Top and More Detailed Formats Below

This system does not allow users to explicitly access an overview picture; there is no 'see all' button. The screen size also means that the user cannot see the entire area within their boundary effectively.

The System should Allow Ease of Navigation Across Displays at the Same Level in the Hierarchy

The refresh rate of the system is too slow to allow for ease of navigation across displays. The time taken in both panning and zooming significantly affects task flow and potentially leads to users becoming disorientated. The system refresh rate should be fast enough to allow the user to navigate easily between displays allowing them to frequently zoom in and out in order to gain environmental perspective.

The System should Permit Users to Switch Between Single and Multiple Window Views

The system does allow users to switch between single and multiple window views. The system has a 'Microsoft Windows'-like environment which allows windows to be overlaid on top of one another. The system does initially minimise the old screen but this screen can easily be maximised and resized to allow the view of more than one screen. Figure 8.5 shows the system displaying multiple window views.

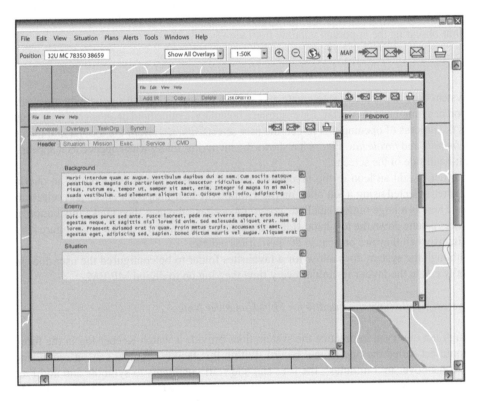

Figure 8.5   Diagram showing overlapping resizable windows and multiple windows displayed on a single screen

The System should Limit the Maximum Number of Windows to be Opened, to Prevent Confusion

The system does not have a default maximum number of windows per screen; there is no limit to the number of windows which can be open simultaneously. It is down to individual users to manage their displays. Users should be encouraged to limit the number of windows open at any one time as this has a significant effect on the terminal processing performance.

The System should Allow Users to Save a Default Configuration of Windows that Suits Their Purposes

There is no automatic default configuration for the terminal. The user does have the ability to develop 'favourite pallets', however these have to be reloaded upon each login. The system should allow the user to configure the windows under their personal log on, so that every time that they log on to the system, their preferred windows layout automatically appears.

The System should Allow Important Windows Always to be on View

As Figure 8.2 demonstrates the system does not ensure that the important windows are always on display. The window displaying the user's LOP is the most important window and it is not viewable in Figure 8.2. The user can easily minimise important windows, or place different windows on top of them thus obscuring the important window from view.

The System should Allow Users to Configure Their Favourites and Auick Access Buttons

The system does allow the user to create a favourites palette. Within this palette the user can place any icon or symbol that they draw, for example, a boundary, a mine field or an enemy tank. Figure 8.6 shows the process of opening this folder, the user selects *situation* from the toolbar and then selects *create/modify* and *favourites*. Figure 8.6 shows the favourites folder when opened as it appears down the left-hand side of the screen.

In order to add an icon to the favourite's palette the user creates the icon on an overlay, the icon is then selected using a left click on the symbol, the user then right clicks and selects the *edit* option. There is a large button labelled '*add to favourites*', the user then clicks this button and the favourites palette down the left-hand side of the screen is populated with the icon. Figure 8.6 shows these items when they are present in the favourites palette.

Although the system does allow for a favourites folder to be configured the user does have to manually create the favourites folder every time they log on to digital MP/BM.

The System should have the Facility for Shift Handover Notes

Although not observed in practice the system does provide a Watch Keeper log in the form of an electronic list called the operational record. Within this facility the user can set reminders, such as 'shift handover in 5 minutes', the user can add staff comments to symbols and add Free Text notations to the operational record. Figure 8.7 shows the route to the operational record. The user is then presented with the operational record as seen in Figure 8.7, which can be populated with details of the current situation.

The System should Allow the User to Configure a List of Displays that will Cycle Through Automatically

This is not possible with the current system. There is little perceived benefit from the addition of this functionality.

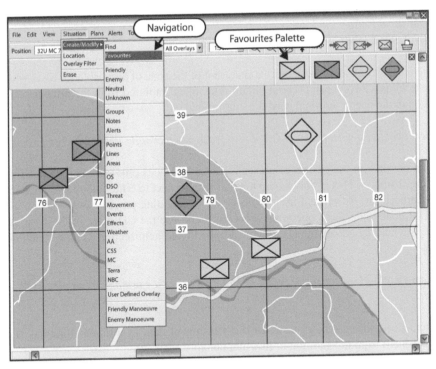

Figure 8.6   Diagram showing how to navigate to the favourites palette

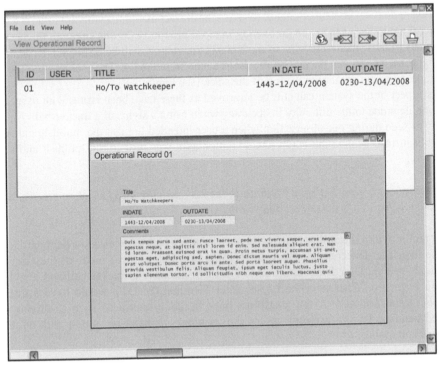

Figure 8.7   Operational record within the Watch Keeper log

The System should Allow the User to Capture a Snap Shot of the State of a Variable, for Discussion Later (stored on a hard drive or printed)

This system does allow users to capture a snap shot of the state of a variable for discussion later as all of the terminals have a print screen button. The image captured from the print screen can then be saved on to the user's terminal hard drive or printed out.

The System should have Hotlinks to Databases, such as Operating Procedures

This is not possible in the system at present. The system does provide a help manual if the user presses the F1 button. This is only a help manual with respect to buttonology and does not include operating procedures. At present in order to view the operating procedures the user must acquire a CD ROM containing these operating procedures and use the CD drive of the terminal to read this disk. It would be advantageous to have a hyperlinked help section in both functionality and doctrine.

The System should Allow for the Viewing of Data in Different Formats (trend plots, site schematics)

The system does not currently allow for the viewing of data in different formats. There is no perceived benefit from the addition of this functionality.

The System should Allow Quick and Unambiguous Selection of the Required Item

The software does support functionality to select the correct mapping symbol. Figure 8.8 shows an example of a number of icons on an overlay in close proximity. In this situation the user can left click on any of these icons and a list of the icons in the vicinity will be displayed (as demonstrated in the figure). The user can then select the desired icon from the list. The user also has the capacity to bring the pertinent symbol to the top of the stack (see Figure 8.8).

This aspect of the system can still be improved as there have been issues with users selecting the wrong item due to the difficulty in accuracy when using a stylus or a tracker-ball. An example of this was seen during our observation when a user intended to push the 'publish and subscribe' button but instead pushed the 'CONTACT' button as it was next to the 'publish and subscribe' button.

The System should Allow Different Methods to Interact with the Interface (for example, mouse, keyboard, touchscreen, tracker-ball and so on)

The system provides the ability to use a mouse, a tracker-ball, a stylus and a keyboard to interact with the interface. Different terminals are limited to which of these methods they can use. For example, a Vehicle User Data Terminal (VUDT) would have a tracker-ball and a stylus along with the keyboard. Whereas a Staff User Data Terminal (SUDT), would have a tracker-ball, a keyboard and a mouse. A few platforms have bespoke interfaces such as thumb joysticks on fixed handles. Due to usability issues the terminal has been retrofitted with an additional keyboard and mouse.

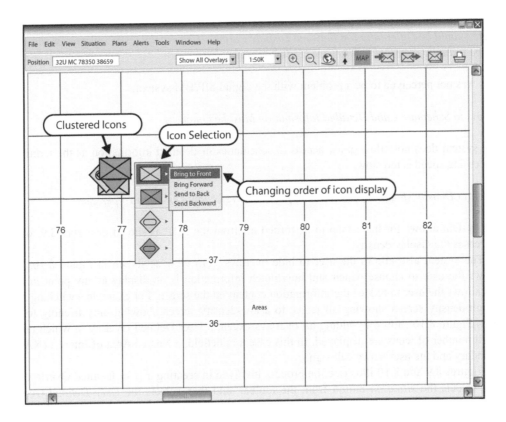

Figure 8.8   Diagram displaying a number of icons on top of one another

The System should Give Quick, Consistent and Effective Feedback to the User

The refresh rate and processing rate of the system is too slow to enable quick, consistent and effective feedback to the user. Very occasionally and inconsistently the system may display an hour glass figure when an action is taking an extended period of time. If a user attempts to open a large product (1MB or greater; for example, an Op Order), a progress bar will appear on the screen. A progress report may also appear if the user sends a large message. The system's feedback is inconsistent and thus unreliable.

There should be Dedicated Displays for Emergency Shut Down Operations

The system does not allow for dedicated displays for emergency shut down operations as it is a single screen system.

There should be Back-up for Loss of Power (batteries and generators)

The terminals' batteries are kept charged by running the vehicle engines for between 15 and 30 minutes every hour. Although this ensures no loss of power, it creates a lot of noise. Static running of the engines at low load is also detrimental to the vehicles.

There should be Differentiation Between Display Formats (for example, site schematics should not be virtually identical – otherwise confusion may occur)

This was not perceived to be a problem with the digital MP/BM system.

Access to Schematics and Detailed Information Must be Quick

The system does not allow quick access to schematics or detailed information as the terminal's processing speed is too slow.

Displays Density should be Less than 50 per cent, Preferably Less than 25 per cent

The system allows for the e-map to be turned off from the LOP as shown in Figure 8.9, which decreases the display density.

   The system also allows the user to hide overlays from view as shown in Figure 8.10. This allows the user to choose which and how much information is on display at any point in time and allows the user to reduce the information content on the screen. For example, switching from a high-density screen showing all icons, to a low-density screen showing only friendly forces, or as Figure 8.10 shows, by hiding all overlays except a User Defined Overlay in which only a small number of icons are displayed. In this case a minefield, a Named Area of Interest (NAI), a boundary and the user's own call sign.

   Figures 8.9 and 8.10 illustrate the process involved in creating a User Defined Overlay. The user selects the *situation* button from the toolbar, and then selects the *User Defined Overlay* option. At this point a palette similar to the 'favourite's palette' is created down the left-hand side of the screen. The palette is populated in the same manner as the 'favourite's palette' shown previously.

   A careful balance needs to be struck between reducing the need for navigation between and within pages. The selection and level of applicability of the information on the screen is far more important that the quantity.

Displays should Not be Cluttered

The system allows for icons to be hidden from view in order to ensure that the display is not cluttered. Icons can be hidden from view in a number of ways in order to ensure that only the relevant information remains on the screen.

Spatial Alignment on Site Schematics should Give a Good Appearance

This requirement is not applicable to this system.

It should Not be Necessary to have More than One Display on View to Perform a Specific Task

The system requires the user to flick between windows, as well as copy and paste between windows in order to perform most tasks.

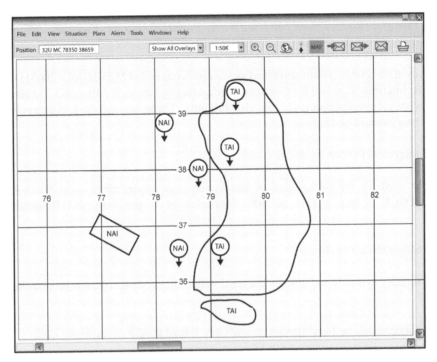

Figure 8.9   Diagram showing the LOP with the e-map turned off

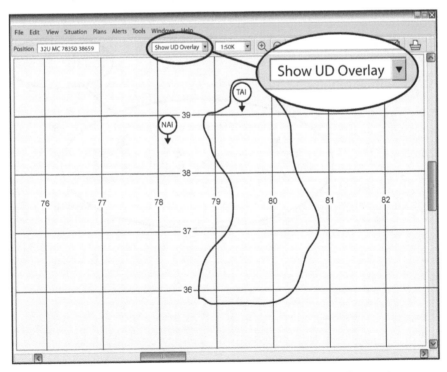

Figure 8.10   Diagram displaying the ability to hide all icons except the user's own

Concrete Pictorial Displays should be Used Where Possible (abstract shapes are difficult to learn)

The icons used within this system are taken from a set of icons called 2525B (DoD (2005) Department of Defense, Interface Standard for Common Warfighting Symbology, MIL-STD-2525B, available at http://www.mapsymbs.com/ms2525b_ch1_full.pdf). These are standard military icons with which all users should be familiar.

A Cool Background Colour should be Used

The colour used for the background of this system can be described using the hexadecimal code #D4D0C8; it has a Hue of 40°, Saturation of 6 per cent and Brightness of 83 per cent.

Garish Colours should be Avoided

Standard military symbols and colours are used within this system. There is no gratuitous use of colour.

Colour Coding should be Used Sparingly, and Only Where it Adds Value to the Tasks

Within this system there is very little colour coding. Aspects of the colour coding that is present are inconsistent in places and this can be confusing for the user. As seen in Figure 8.11

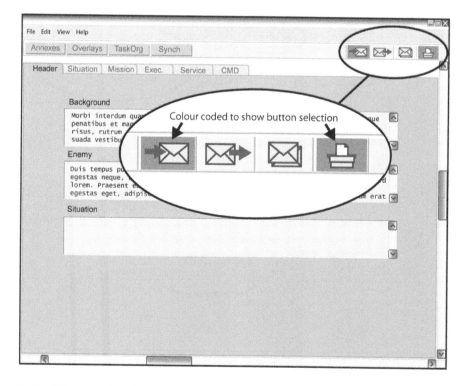

Figure 8.11   Diagram showing the purple colour coding of certain buttons

a number of buttons within the system are colour coded purple. Purple colour coding appears to carry the message that if the user clicks this then something will happen straight away. The colour coding acts as a form of warning message; it prompts the user to consider their actions before clicking the button. The colour coding of buttons is used correctly for publish and subscribe, and send. However, the purple colour coding is also used inconsistently for the print button; the user is presented with a new window asking them to select a printer, this window can be cancelled. The print button should be grey and not purple as it does not lead directly to an action.

Within the system a further colour coding system is used to show when a button has been switched on. When a button has been switched on it will turn green to alert its state to the user (see the menu bar in Figure 8.11)

Animated Graphics should be Used Sparingly. Only Use Where it is Really Necessary to Attract Attention

There are no animated graphics within the digital MP/BM system.

Summary

Of the 28 applicable requirements outlined by EEMUA 201:2002 the digital MP/BM system examined was: fully compatible with 8; partially compatible with 12; and incompatible with 8 (see Table 8.1 on the following pages). As these requirements do not carry equal weighting it is not appropriate to draw wider conclusions from this summary. Rather, the information should be used to inform future software and hardware upgrades.

Table 8.1   Summary of EEMUA survey

		Fully compatible	Partially compatible	Incompatible	Not Applicable
1	The number of screens must allow for complete access to all necessary information under all operational circumstances			X	
2	The design should allow for a permanently viewable overview display format		X		
3	Continuous access to alarm information should be provided				X
4	The capability to expand the number of screens should be built into the design		X		
5	The provision of screens should allow for start-up, steady state monitoring, shut-down and abnormal situations		X		
6	The system should allow the user to resize windows easily	X			
7	The system should enable ease of navigation up and down the hierarchy of screens		X		
8	The system should have a simple hierarchical structure, with a total plant overview at the top and more detailed formats below		X		
9	The system should allow ease of navigation across displays at the same level in the hierarchy		X		
10	The system should permit users to switch between single and multiple window views	X			
11	The system should limit the maximum number of windows to be opened, to prevent confusion			X	
12	The system should allow users to save a default configuration of windows that suits their purposes		X		
13	The system should allow important windows always to be on view			X	
14	The system should allow users to configure their favourites and quick access buttons		X		
15	The system should have the facility for shift hand-over notes		X		
16	The system should allow the user to configure a list of displays that will cycle through automatically				X
17	The system should allow the user to capture a snap shot of the state of a variable, for discussion later (stored on a hard drive or printed)	X			
18	The system should have hotlinks to databases, such as operating procedures			X	

Table 8.1   *Concluded*

		Fully compatible	Partially compatible	Incompatible	Not Applicable
19	The system should allow for the viewing of data in different formats (trend plots, site schematics)				x
20	The system should allow quick and unambiguous selection of the required item		x		
21	The system should allow different methods to interact with the interface (e.g., mouse, keyboard, touchscreen, tracker-ball, etc)		x		
22	The system should give quick, consistent and effective feedback to the user			x	
23	There should be dedicated displays for emergency shut down operations				x
24	There should be back-up for loss of power (batteries and generators)	x			
25	There should be differentiation between display formats (e.g., site schematics should not be virtually identical – otherwise confusion may occur)				x
26	Access to schematics and detailed information must be quick			x	
27	Displays density should be less than 50%, preferably less than 25%				x
28	Displays should not be cluttered			x	
29	Spatial alignment on site schematics should give a good appearance				x
30	It should not be necessary to have more than one display on view to perform as specific tasks			x	
31	Concrete pictorial displays should be used where possible (abstract shapes are difficult to learn)	x			
32	A cool background colour should be used	x			
33	Garish colours should be avoided	x			
34	Colour coding should be used sparingly, and only where it adds value to the tasks		x		
35	Animated graphics should be used sparingly. Only use where it is really necessary to attract attention	x			
		8	12	8	7

Chapter 9
Usability Questionnaire

Context

The effectiveness of a tool is dependent upon its appropriate use. The extent to which the tool will be utilised appropriately is dependent upon how easy and intuitive the tool is to use (its usability). When the tool in question is computer software, the mechanics of the tool are hidden from the user and the means of Human Computer Interaction (HCI) is provided in the form of an 'interface'. The primary options for facilitating knowledge transfer on the use of a software tool are user training, the availability of Subject Matter Experts (SMEs) or providing a user interface which overtly encompasses its purpose, navigation and method of operation.

In situations of low stress, where errors in tool use have low associated risks and where training and SMEs are freely available to the user, the usability of the interface (while still highly desirable from an efficiency point of use) is not critical. In situations of high stress, where errors in use have high associated risks and where the time and resources available do not allow users to take advantage of training and SMEs, the usability of the interface is *absolutely critical* to the effectiveness of the tool. Live digital Mission Planning and Battlespace Management (MP/BM) falls into the latter category and as such the usability of the HCI provided by the digital MP/BM tool is critical to its effectiveness.

Method

To determine the usability of HCIs, a simple and powerful method of assessment is a subjective user-focused 'usability checklist'. Subjective checklists rely on the users of the system to provide their personal opinion within a structured controlled approach, in this case a questionnaire. All subjective opinion should be considered in context; however, as part of a wider data collection approach this analysis technique provides an invaluable insight into the opinions of the users flagging up areas of critical concern within the system.

For this study the most appropriate checklist for evaluating the digital MP/BM software tool is Ravden and Johnson's (1989) HCI checklist. This checklist comprises ten sections of questions that assess the overall usability of a particular system. These sections are outlined below, with Ravden and Johnson's (1989) recommendations for successful usability highlighted.

1. *Visual clarity.* Refers to how clearly the system displays information. **Information displayed on the screen should be clear, well organised, unambiguous and easy to read.**
2. *Consistency.* Refers to how consistent the interface is in terms of appearance, the presentation of information and the ways in which users perform tasks. **The way the system looks and works should be consistent at all times.**

3. *Compatibility*. Refers to how compatible the system is with other related systems. **The way the system looks and works should be compatible with user conventions and expectations.**

4. *Informative feedback*. Refers to the level, clarity and appropriateness of the feedback provided by the system. **Users should be given clear, informative feedback on where they are in the system, what actions they have taken, whether these actions have been successful and what actions should be taken next.**

5. *Explicitness*. Refers to the clarity with which the system conveys its functionality, structure and capability. **The way the system works and is structured should be clear to the user.**

6. *Appropriate functionality*. Refers to how appropriately the system functions in relation to the activities that it is used for. **The system should meet the needs and requirements of users when carrying out tasks.**

7. *Flexibility and control*. Refers to the flexibility of the system and the level of control that the user has over the system. **The interface should be sufficiently flexible in structure, in the way information is presented and in terms of what the user can do, to suit the needs and requirements of all users and to allow them to feel in control of the system.**

8. *Error prevention and correction*. Refers to how well the system prevents user errors from being made, and to what extent user's errors are allowed to impact task performance (error tolerance). **Systems should minimise the possibility of user error and possess facilities for detecting and handling the errors that do occur. Users should be able to check inputs and correct errors or potential error situations before inputs are processed.**

9. *User guidance and support*. Refers to the level of guidance and support that the system provides to its end users. **Systems should be informative, easy-to-use and provide relevant guidance and support, both on the system itself (for example, help function) and in document form (for example, user guide), to help the user understand and use the system.**

10. *System usability*. Refers to the overall usability of the system.

During the trial, a questionnaire was presented to all soldiers at Brigade (Bde) Headquarters (HQ) and Battle Group (BG) HQ who came into contact with the digital MP/BM system. The questionnaire comprised 140 questions, separated into the ten sections described above. A 'tick the box' format was used, with an additional section at the end for comments on usability.

Twenty-six questionnaires were returned from Bde HQ and 13 from BG HQ. The data was analysed both overall and by group to see if differences were experienced in usability.

Results

The format of the questionnaire provided between five and 25 questions per section. Radar plots were used to succinctly represent the impact of each section. Each axis illustrates how the participants rated each question: positive responses are represented at the perimeter, negative responses at the centre. Median values were used to remove the impact of inaccuracies cause by isolated outlying responses.

Visual Clarity

Figure 9.1 shows the overall median values for Visual Clarity. The results show that participants found only 'some of the time' that it was:

'Easy to find information, schematics were clear, screens were uncluttered, information was logically organised, information clears before updating and there was colour blind suitability.'

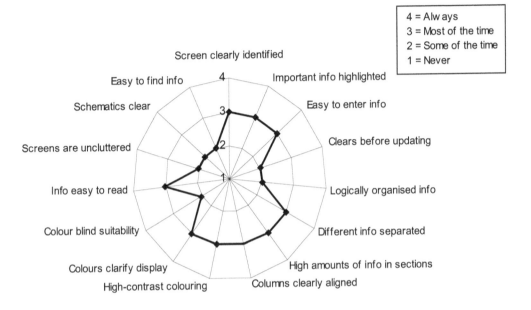

Figure 9.1   Overall median values for Visual Clarity

All other factors were experienced 'most of the time'. The general experience is summed up by the following participant comment:

'The digital MP/BM screen always looks busy. The windows that include written information are generally too small and it is often time consuming to find the information you want.'

In addition, there is frustration that information does not clear before updating:

'Forms do not default to empty. Once filled, the user needs to clear the form before using again to prevent misinformation.'

Figure 9.2 compares the median values for Visual Clarity for Bde HQ and BG HQ. Here we can see there is clear disparity between the two groups in five factors. Bde HQ gave far more positive answers, responding that:

'Screens were clearly identified, info is easy to read, columns are clearly aligned and high amounts of info are separated into sections *'most of the time'*

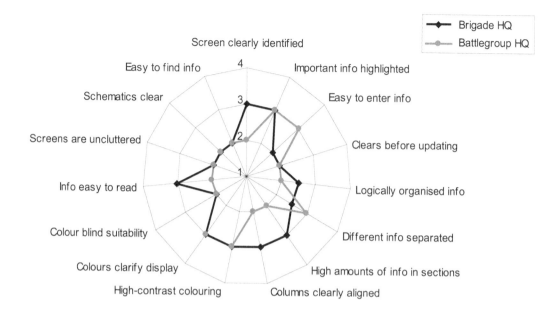

Figure 9.2   Comparison of median values for Visual Clarity by group

Whilst BG HQ experienced these factors only 'some of the time'. Conversely, BG HQ found it was *easy to enter information* 'most of the time' whereas Bde HQ found this only 'some of the time'.

Part of the reason for this difference could be the fact that participants in BG HQ had not had exposure to previous versions of digital MP/BM. Their first impressions, therefore, may represent a more accurate portrayal of the visual clarity of the software. Since participants from Bde HQ were familiar with previous versions, we would expect them to elicit a more positive experience where improvements since the last version were apparent.

From Ravden and Johnson's (1989) recommendations for Visual Clarity, the results show that *information displayed on the screen is clear and easy to read* 'most of the time', but *well organised and unambiguous* only 'some of the time'.

Consistency

Figure 9.3 shows the overall median values for Consistency. Here we can see participants felt they *experienced all factors* 'most of the time'. Whilst this is a positive response for the digital MP/BM software tool, participant comments expressed a lack of consistency with other software packages they use:

> 'All the above are consistent in digital MP/BM but not similar to windows, – this is the operating system that the majority of people are familiar with.'

By not making use of participants' existing skills with external software, users cannot transfer their knowledge to the new system, increasing the learning curve and frustrations as they apply an incorrect 'mental model' when using the system.

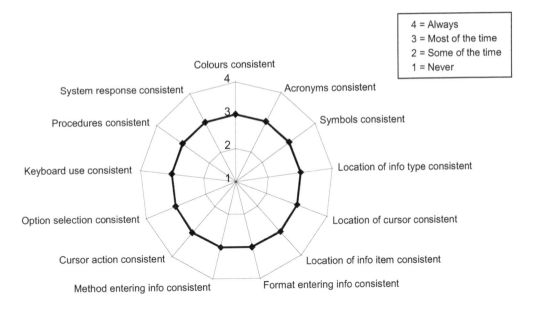

Figure 9.3   Overall median values for Consistency

Figure 9.4 compares the median values for Consistency for Bde HQ and BG HQ. Here we can see two main areas of disparity. Where Bde HQ experience that *system response is consistent and location of cursor is consistent* 'most of the time', BG HQ only experience these factors 'some of the time'. There is a similar, although less marked, difference when considering if *methods of entering info are consistent*, where Bde HQ is more positive. For the category acronyms that are consistent, BG HQ do experience a slightly more positively experience than Bde HQ.

This difference is again likely to be related to the different levels of experience in using the digital MP/BM software. When considering the consistency of system response, comments from BG HQ relate more to the processing speed of the system being inconsistent, rather than the software itself.

Ravden and Johnson (1989) recommend that the way the system looks and works should be consistent *at all times*. The results suggest that digital MP/BM meets these requirements *most of the time*.

Compatibility

Figure 9.5 shows the overall median values for Compatibility. Participants experienced only 'some of the time' that *the system works as the user expects, there is an expected sequence of activities, the information fits the user's view of the task* and *control actions are compatible with those used in other systems with which the user may need to interact*. All other factors are experienced 'most of the time'.

Participant comments again echo that the lack of similarity to products with which they are already familiar hampers their use of the system as they are applying an inappropriate mental model:

'Majority of users are competent on Microsoft, this doesn't look or feel like Microsoft and therefore appears over complex.'

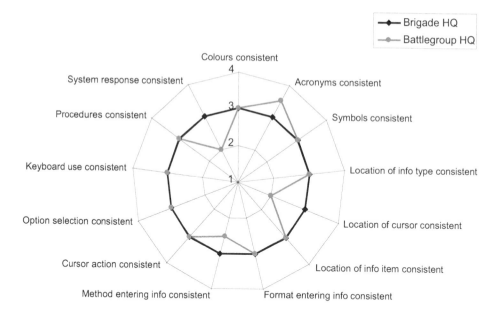

Figure 9.4   Comparison of median values for Consistency by group

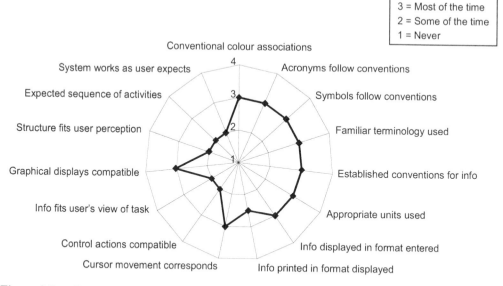

Figure 9.5   Overall median values for Compatibility

'Digital MP/BM is not intuitive. An advanced knowledge of the system is required for the user to be proficient.'

Figure 9.6 compares the median values for Compatibility for Bde HQ and BG HQ. Experience of the system varies most between groups when considering if the *structure of information fits the user's perception of the task*, if *familiar terminology is used* and if *the format and sequence*

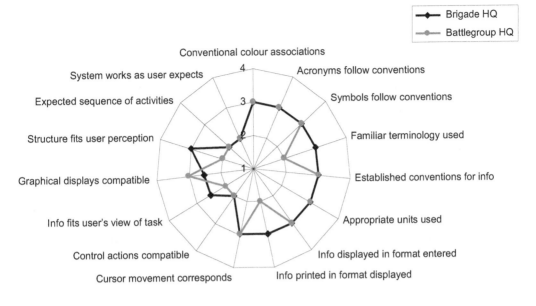

Figure 9.6 Comparison of median values for Compatibility by group

in which information is printed is compatible with the way it is displayed on the screen. Bde HQ experiences these factors 'most of the time' whilst BG HQ encounters these only 'some of the time'.

Whilst some of this can again be put down to lack of familiarity with the system, BG HQ's poorer rating for *information is printed in the format displayed on the screen* can possibly be attributed to their reduced facilities and resources when compared to Bde HQ. BG HQ have the use of an A4 printer, which is unable to spool large print jobs and cannot cope with the output (which often defaults to A3 or larger). The I-Hub team is made up of one Information Manager, so there is less opportunity to find workarounds to achieve their printing goals. Both Bde HQ and BG HQ feel the print function is overcomplicated, difficult to set up and does not produce the expected output.

Ravden and Johnson (1989) recommend that the way the system looks and works should be compatible with user conventions and expectations. Whilst digital MP/BM is compatible most of the time when displaying symbology and data, it falls short in compatibility with users' perceptions and expectations.

Informative Feedback

Figure 9.7 shows the overall median values for Informative Feedback. Here we can see only two factors are represented positively, *with concise instructions (instructions and messages displayed by the system are concise and positive)* and *required information to be entered is clear* being experienced 'most of the time'. All other factors are experienced only 'some of the time'.

Comments from users mirror the lack of confidence in their actions as a result of poor informative feedback:

'The system has been reset several times by users thinking it has "hanged" when in fact it is still processing.'

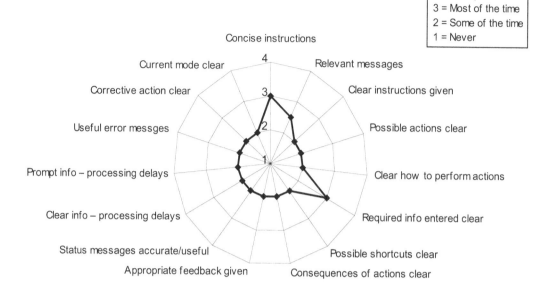

Figure 9.7   Overall values for Informative Feedback

'You are never sure if you have done the right thing.'

A particular frustration is the process whereby to complete an action, users must close the screen. Most users are familiar with 'OK' or 'apply' buttons found in other software packages which provide reassurance that the action has been accepted. In addition, status messages get lost when users move to a new window, meaning they must leave additional windows open to keep track. This in turn slows down processing speed and increases the screen clutter.

Figure 9.8 compares the median values for Informative Feedback for Bde HQ and BG HQ. In three instances we can see that Bde HQ are more positive than BG HQ, experiencing *concise instructions*, *relevant messages* and *clear instructions* are given 'most of the time', whereas BG HQ experiences this only 'some of the time'. Conversely BG HQ are more positive than Bde HQ when considering if *the required information to be entered on a screen is clear* and if *the corrective action necessary to correct errors is clear*. For the former, BG HQ experiences this 'most of the time' compared to Bde HQ who find this only 'some of the time'. For the latter, BG HQ only experience this 'some of the time' and Bde HQ 'never' experience clear corrective action.

Bde HQ's slightly more positive response relates to factors that would improve with familiarity. The biggest issue here for both groups is the lack of clear corrective action for errors, causing great user frustration, as expressed in the following comment:

'They tell you something is wrong. Not what, where or how to fix it. For example, antenna alarm when there are 5 antennas!'

Whilst digital MP/BM does fulfil one aspect of Ravden and Johnson's (1989) recommendations for Informative Feedback, in that most of the time users are given clear, informative feedback on

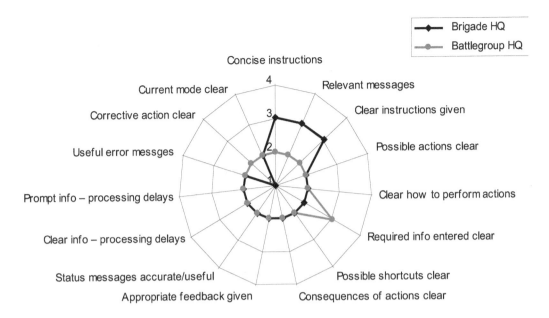

Figure 9.8   Comparison of median values for Informative Feedback by group

where they are in the system, it is inadequate at giving users feedback on what actions they have taken, whether these actions have been successful and what actions should be taken next.

Explicitness

Figure 9.9 shows the overall median values for Explicitness. The results show only one factor, *it is clear what part of the system the user is in*, is experienced 'most of the time'. All other factors occur only 'some of the time', which give an overall negative result for explicitness. To sum up the participants experience:

'Digital MP/BM is complex, unintuitive and difficult to learn.'

If the way the system works and is structured is not clear to the user, it necessarily requires a great deal of mental effort to achieve the desired goal. Skill fade will be high as users have to rely on their memory rather than an obvious and logical Course of Action (CoA) presented by the interface. In high stress situations, this is likely to lead to frustration and errors.

Figure 9.10 compares the median values for Explicitness for Bde HQ and BG HQ. Here we can see that Bde HQ are considerably more positive than BG HQ in two areas: Where Bde HQ feel it is clear what part of the system the user is in 'most of the time', BG HQ experience this only 'some of the time. Where Bde HQ considers that there is an obvious system structure only 'some of the time', BG HQ feels they 'never' experience this.

Again, Bde HQ's greater familiarity with the digital MP/BM system is likely to have provided slightly more positive results. Nevertheless, both groups give predominantly negative responses.

Ravden and Johnson (1989) recommend that the way the system works and is structured should be clear to the user. The results show that digital MP/BM undoubtedly falls short in this category.

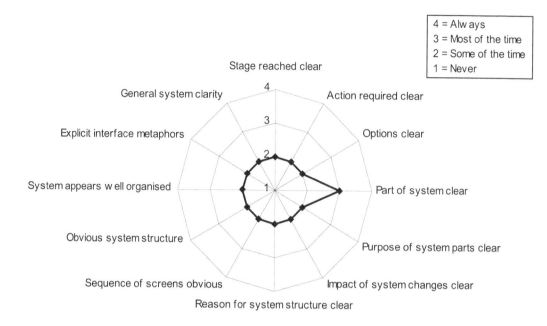

Figure 9.9   Overall median values for Explicitness

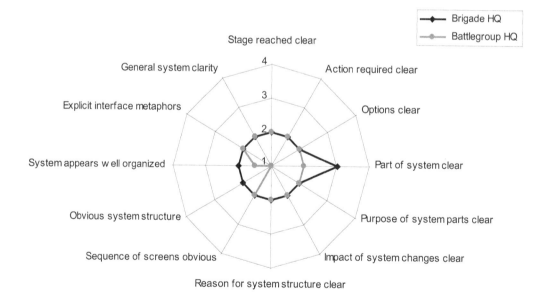

Figure 9.10   Comparison of median values for Explicitness

Appropriate Functionality

Figure 9.11 shows the overall median values for Appropriate Functionality. All factors are shown to score poorly as they are experienced only 'some of the time'.

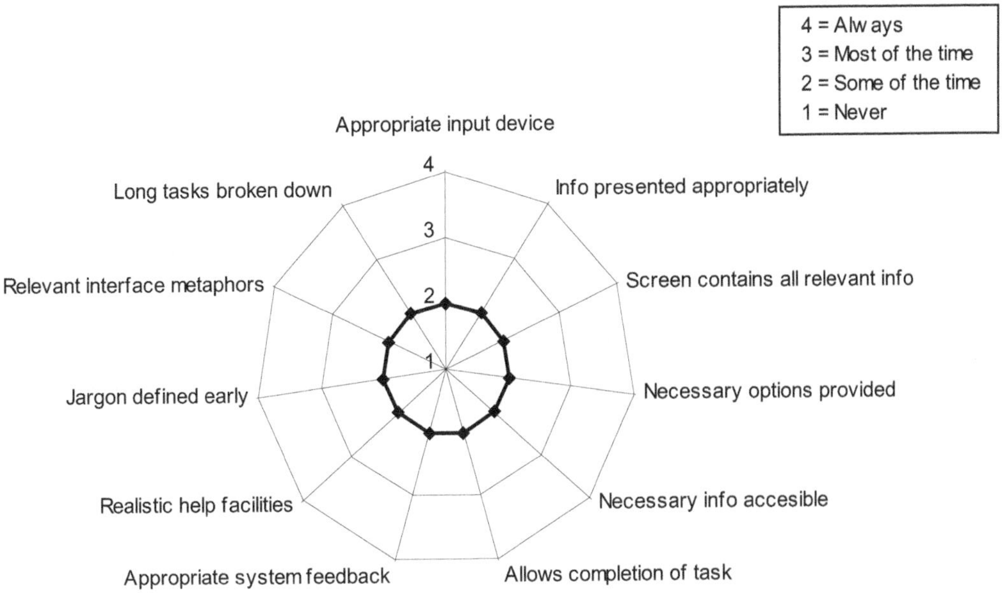

Figure 9.11 Overall median values for Appropriate Functionality

The greatest frustration was found with the input device. The thumb mouse and keyboards provided were very rigid, severely hindering the users' ability to navigate to the desired button, or touch type text.

There are also further interrelated issues with hardware when trying to complete a task, as illustrated by the following comment:

'The user needs to have multiple screens up to carry out most of the functions. This puts strain on the processor causing frequent crashes.'

This highlights that effective use of the software tool is highly dependent on the correctly designed software being capable of running on the available hardware. During the exercise, the digital MP/BM software and the hardware supplied showed evidence of incompatibility.

Figure 9.12 compares the median values for Appropriate Functionality for Bde HQ and BG HQ. The results show the only discrepancy between the two groups is found with *the input device available is appropriate for the tasks to be carried out*. Where Bde HQ encounter this only 'some of the time', BG HQ feel they 'never' experience this.

There is very little difference in the results for the two groups in terms of Appropriate Functionality. The more positive results at Bde HQ are likely to reflect that external mouse and keyboards had been provided for these users by the time the questionnaire was completed. BG HQ did not receive this additional hardware during the trial.

Ravden and Johnson (1989) recommend that the system should meet the needs and requirements of users when carrying out tasks. The results above show that digital MP/BM only achieved this some of the time.

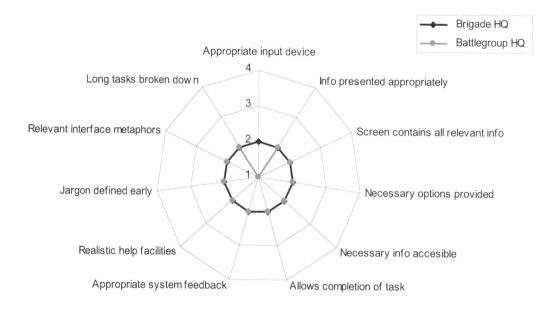

Figure 9.12   Comparison of median values for Appropriate Functionality by group

Flexibility and Control

Figure 9.13 shows the overall median values for Flexibility and Control. Whilst most of the factors score poorly, we can see positive responses in three areas: *custom file naming, main menu is always accessible* and *different parts of the system are accessible as required* are experienced 'most of the time'.

Comments from participants express a great need for an 'undo' or 'back' button. This would give them more confidence when entering information and performing actions.

It should also be noted that even though there are positive comments relating to accessing the main menu, users have expressed that this takes between 3 to 5 minutes due to the processing speed, again showing the interdependence between software and hardware in terms of user experience.

Figure 9.14 compares the median values for Flexibility and Control for Bde HQ and BG HQ. BG HQ experience *custom interface preferences* and *main menu accessibility* 'most of the time', compared to Bde HQ who felt these were represented only 'some of the time'. However, Bde HQ did have a more positive experience than BG HQ with regards to accessing different parts of the system. Both groups are still predominantly negative and from the comments provided there do not appear to be clear reasons for the differences in their experiences.

Ravden and Johnson (1989) recommend that the interface should be sufficiently flexible in structure, in the way information is presented and in terms of what the user can do, to suit the needs and requirements of all users and to allow them to feel in control of the system. The results above show that digital MP/BM only achieved this some of the time, with the exception of system navigation and custom file naming, where users felt they had control most of the time.

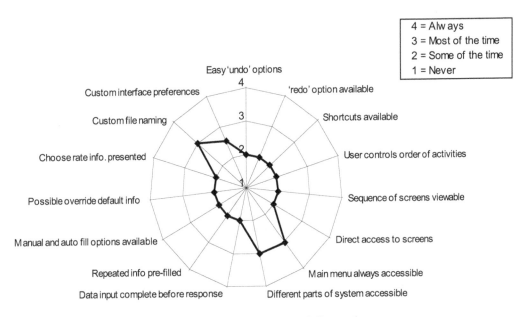

Figure 9.13   Overall median values for Flexibility and Control

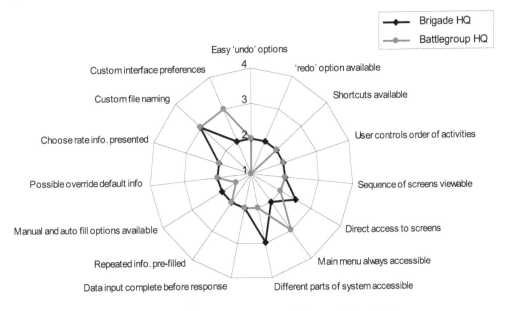

Figure 9.14   Comparison of median values for Flexibility and Control by group

Error Protection and Correction

Figure 9.15 shows the overall median values for Error Protection and Correction. Again the results show low ratings with all but two of the factors experienced only 'some of the time'. Participants did feel the system *prevents unauthorised actions* and that they could check *before their actions were processed*, 'most of the time', however.

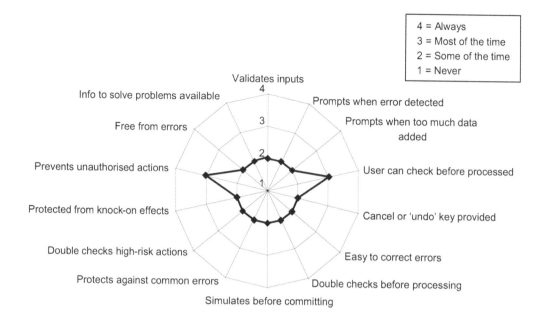

Figure 9.15 Overall values for Error Prevention and Correction

The general feeling from user comments is that not only is it difficult to correct errors, but that the system interface actually encourages errors.

'Especially vulnerable to common errors due to the complexity of simple tasks.'

'I've barely been able to correct digital MP/BM errors myself – sometimes it has beaten the local experts too.'

Figure 9.16 compares the median values for Error Prevention and Correction for Bde HQ and BG HQ. BG HQ is more positive than Bde HQ in two areas, experiencing that 'most of the time' the system *prompts the user when an error is detected* and *double checks high-risk action*s. BG HQ is more negative than Bde HQ in feeling that the system is 'never' *free from errors*, or *simulates before committing an action.*

It is interesting to note that whilst BG HQ response to error prompts is that they are *'often too frequently interfering with ongoing work'*, Bde HQ feels that *'mostly it does not prompt'*.

It is not clear whether the greater familiarity with the system has led Bde HQ's to 'ignore' irrelevant error prompts, or that BG HQ's less experienced use of the system causes more errors, resulting in more error messages.

The results show that digital MP/BM partially supports Ravden and Johnson's (1989) recommendations for Error Prevention and Correction. Whilst only some of the time, the system minimises the possibility of user error and possesses the facilities for detecting and handling the errors that do occur; most of the time the user *is* able to check inputs and correct errors before inputs are processed.

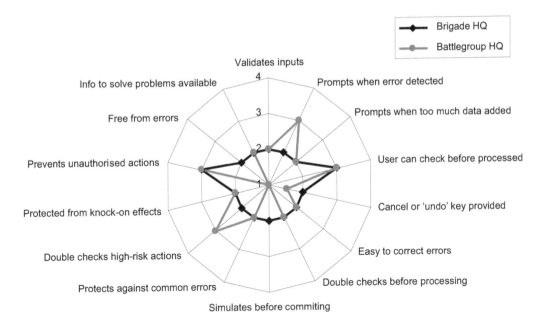

Figure 9.16   Comparison of median values for Error Prevention and Correction by group

User Guidance and Support

Figure 9.17 shows the overall median values for User Guidance and Support. The results show all factors scored poorly by being present only 'some of the time'. There seemed to be a general lack of understanding of the help that the system could provide, and a number of different paper-based documents providing contradictory workarounds. Time constraints and problems finding the required information led users to seek help from a specialist rather than trying to solve the problem themselves.

> 'For the exercise many aid memories were provided by different organisations – often conflicting.'

> 'More often an application specialist will be called to advise rather than sitting through a help file system which if printed out is 960 pages of A4 in size.'

Figure 9.18 compares the median values for User Guidance and Support for Bde HQ and BG HQ. Here we can see that BG HQ felt better supported than Bde HQ, believing that *user guidance was current* and *an online help facility was available* 'most of the time'.

The results here are slightly misleading since the comment from BG HQ suggests that they have not used the online help facility, but assume that this exists. Because of this, the positive result that the user guidance is current loses some of its validity.

Ravden and Johnson (1989) recommend that systems should be informative, easy-to-use and provide relevant guidance and support, both on the system itself and in document form, to help

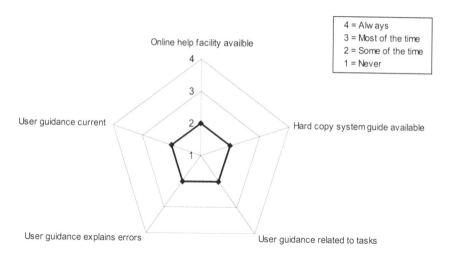

Figure 9.17   Overall values for User Guidance and Support

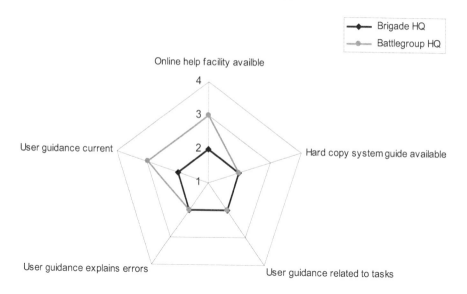

Figure 9.18   Comparison of median values for User Guidance and Support by group

the user understand and use the system. During the trial, the number of workaround documents developed verified that this had not been achieved with digital MP/BM.

System Usability Problems

Figure 9.19 shows the overall median values for System Usability Problems. Out of 25 different factors, six were considered a major problem, 15 a minor problem and only four no problem at all. The 'major problems' centre around errors, awkward input devices and slow response times. The factors which are considered to be 'no problem' are less critical issues such as colours and readability, and given previous criticism of the processor speed, the unlikely issue of response

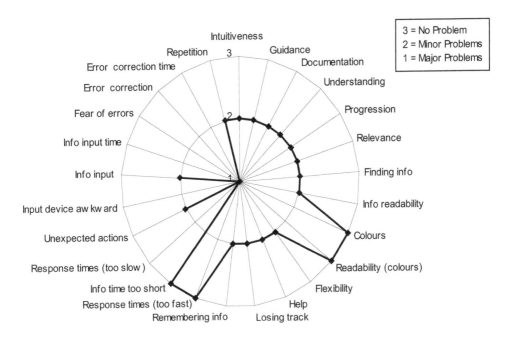

Figure 9.19 Overall median values for System Usability Problems

times and information times being too fast for the user. All other factors are considered a 'minor problem'.

The fact that all the error-related factors are listed as a 'major problem' is particularly worrying in a system with high associated risks.

Figure 9.20 compares the median values for System Usability Problems for Bde HQ and BG HQ. From these results we can see that there are only minor differences (at most half a point) in each group's experiences.

Conclusion

Figure 9.21 shows users' overall responses for each of the nine categories that according to Ravden and Johnson (1989) influence the usability of a HCI. Here we can see all categories are rated either 'neutral' or 'moderately unsatisfactory'. Rated as moderately unsatisfactory, *explicitness, information feedback,* and *error prevention and correction* will have a negative impact in terms of how efficiently and confidently users will be able to use the system on a day-to-day basis. The mismatch between hardware and software that hinders *appropriate functionality* and the lack of *flexibility and control* causes a great deal of frustration to users. One comment sums up the general feeling that participants have about digital MP/BM:

'The system does what is says on the tin but only with workarounds and very slowly.'

Put another way, it is insufficient to provide the capability without the usability.

Figure 9.22 compares the overall responses for each of the nine categories by group. It is clear that Bde HQ had a more positive experience than BG HQ, however they were still unable to provide a score that was satisfactory. When considering the two groups, the facilities, resources

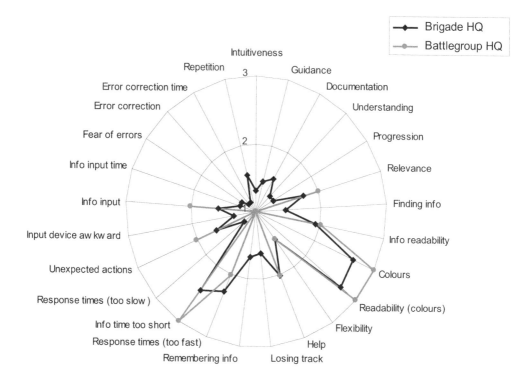

Figure 9.20 Comparison of median values for System Usability Problems by group

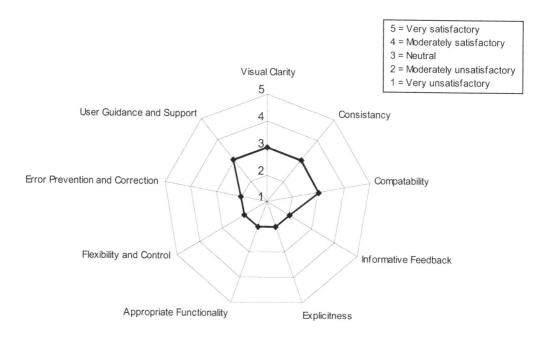

Figure 9.21 Overall median values for categories 1 to 9

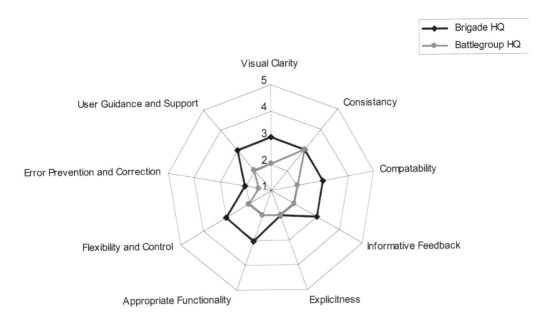

Figure 9.22   Comparison of median values for categories 1 to 9 by group

and familiarity were greater in Bde HQ, providing a lower stress situation than that for BG HQ, all of which are likely to have contributed to the improved experience.

When evaluating the digital MP/BM software tool, the usability of its interface was considered of utmost importance as the tool will be used in situations of high stress, where errors in use have high associated risks and where the time and resources available do not allow users to take advantage of training and SMEs. The results from the usability questionnaire distributed during the exercise reveal that the interface falls considerably short of intuitively encompassing its purpose, navigation and method of operation.

Chapter 10
Environmental Survey

Brief Introduction

The digital MP/BM system resides in a battlefield context, meaning that under a systems perspective it is appropriate to extend the boundaries to encompass the physical conditions that affect operator performance with it. These physical conditions are the province of 'traditional Ergonomics' and cover temperature, air quality, aesthetics, noise and vibration. It is important to note that a full environmental assessment is beyond the scope and resources available for the trial; however, the range of techniques deployed enable major issues to be captured using the minimum of equipment.

In order to contextualise the results, the analysis of the exercise is compared against BS EN ISO 11064-6:2005. This is the British Standard relevant to control room environments and the one that relates most closely to the sort of tasks being performed by military planning personnel. Whilst it may seem incongruous to compare military command and control with what appears to be a sanitised and largely benign civilian counterpart, the following needs to be born in mind:

- Command and control, and indeed NEC, is not confined to the military.
- Despite the rigours of the military setting there still remains an inviolable duty of care.
- Whilst it may be neither possible nor desirable to model military command and control on a civilian alternative, there is value in assessing the gap between what is acknowledged as best practice in other domains.

Best Practice Reference Point

Part 6 of BS EN ISO-11064-6:2005 'Environmental requirements for control centres' deals with the key areas of temperature, light, noise, air quality, vibration and aesthetics in relation to the type of tasks that military planning staff undertake.

BS EN ISO-11064-6:2005 contains general principles for environmental design, requirements and recommendations for each key area and more explicit, informative guidance in terms of actual values to achieve. The spirit of the standard, however, is captured in Table 10.1.

To readers from a military background these guidelines may appear a long way from conditions in Brigade (Bde) and Battle Group (BG) Headquarters (HQ). The point is made again that this is best practice from a human performance point of view and is used merely as a comparison. The sections that follow deal with each environmental factor in turn with an emphasis on comparing the military setting with the guidelines enshrined in BS EN ISO 11064-6:2005.

Data Collection

Environmental assessments were undertaken at both Bde and BG. Nineteen spot measurements were taken at the former and 16 at the latter, spanning the entire Command Post Exercise (CPX) period. To reiterate, a full Environmental Analysis would be considerably more comprehensive but is beyond the remit and resources of the current endeavour. The aim is to provide a rapid,

Table 10.1   Environmental requirements and recommendations from BS EN ISO 11064-6:2005

BS EN ISO 11064-6	DESCRIPTION
Section 5.1: Ambient Temperature	The goal is to achieve a comfortable thermal environment, devoid of localised temperature related anomalies, one over which the operators have some degree of control. Concomitantly, Heating, Ventilating and Air Conditioning (H-VAC) equipment should be fit for the task and able to maintain a stable temperature despite normal variations in external conditions. The thermal environment is not to be specified by air temperature alone but also other factors such as humidity.
Section 5.2: Air Quality	An appropriate cycling rate of fresh outside air into the control room should be provided but not in such a way as to introduce draughts, nor to admit pollutants, particulates and/or stimulate the growth of bacteria. Positive air pressure (or even air tightness) can be beneficial in terms of helping to prevent ingress of pollution, especially in high hazard situations. The H-VAC equipment itself should not introduce vibration whilst grills and vents should be of a type, and installed in such a way, as to facilitate cleaning, draught free airflow and good air circulation.
Section 5.3: Lighting	Lighting conditions are to be designed so that they are appropriate to the visual demands of operators whilst reducing at all times artefacts such as glare. A reference to natural daylight is to be provided and the lighting scheme should incorporate diversity, with a mixture of artificial and natural light (both of which should be operator adjustable). Levels of illumination should, in addition, refer to the 'maintained level' over the lifespan of the lighting device.
Section 5.4: Acoustics	An appropriate acoustic environment should be provided given the operators' task, communication and alarm handling needs. To that end a control room acoustically isolated from noise generating external elements, and one with low reverberation times, should be provided. Consideration should be given to further acoustic isolation of noise generating equipment found within the control room environment itself.
Section 5.5: Vibration	Vibration is, as far as practicable, to be eliminated from the control room environment. This means constructing the control room away from vibration generating equipment, or else isolating the equipment and/or isolating the control room itself. Note that many of the measures for acoustic isolation will be applicable to vibration isolation as well.
Section 5.6: Interior Design and Aesthetics	Generally, the use of low ceilings (or otherwise poorly proportioned rooms), sharp colour contrasts, heavily textured surfaces and a brutalist, overly mechanical look should be avoided. Instead, it is recommended that pale colours be used (lighter ceilings, darker floors) along with appropriate furniture and plants in order to 'humanise' the space. Consideration needs to be given to wear and tear and the implications of 24 hour usage. It is also recommended that operators participate in the design of the control room and in the selection of wall and floor coverings etc.

minimally intrusive assessment in order to capture the major issues in relation to the many other analysis threads connected to digitisation and digital MP/BM.

Ambient Temperature

Minimum requirements

'Ideal' thermal conditions for command and control-type tasks are extracted from BS EN ISO 11064-6:2005 and presented in Table 10.2 alongside the mean values based on live measurements of Bde and BG HQ.

At this summary level of analysis, and notwithstanding the levelling effect of working from averages, there is a clear divergence evident between live military environments and values that

Table 10.2   Comparison of ideal and mean recorded temperature values

EN ISO 11064-6: 2005	Guideline for Winter Conditions	Brigade	Battle Group
Temperature	20–24°C	10.9°C	9.2°C
Vertical Air Temperature Difference	<3°C	5.4°C*	2.5°C*
Surface Temperature of the Floor	19–26°C	8.5°C	6.9°C
Relative Humidity	30–70%	67%	70.9%

* Air temperature asymmetry

are suggested as optimal. The primary difference is of course air and floor temperature, with the military planning context considerably colder than guidelines suggest they should be in order to avoid decrements in human performance. It is further interesting to note a large vertical temperature differential in the relatively more luxurious setting of the Bde HQ. This seems to have arisen because of it being heated more aggressively (using blowers), yet it appears that the larger the indoor/outdoor temperature differential is, the greater the vertical heat loss becomes. In both Bde and BG a temperature differential existed between the main tent and the Warriors, with this difference being particularly pronounced for BG HQ (4.8°C, which is in excess of the recommended limit for vertical temperature differences).

Human Factors Best Practice

The minimum requirements are a relatively crude measure of compliance with standards. What they do not do is map the objective measure of temperature on to the subjectively felt state of being hot or cold (which is the greater determinate of human performance). Neither do they recognise mitigating factors such as clothing and activity level. Human Factors best practice in this respect is the Predicted Mean Vote (PMV) index. This 'predicts the mean value of the overall thermal sensation of a large group' based on estimates of metabolic rate and clothing insulation and estimates or actual measurements of air temperature, mean radiant temperature, relative air velocity and humidity. Based on measurements and/or estimates of these six variables, the PMV value predicts how a large group of people will rate the thermal environment according to the following scale:

+3 = Hot
+2 = Warm
+1 = Slightly Warm
 0 = Neutral
-1 = Slightly Cool
-2 = Cool
-3 = Cold

The goal, generally, is to achieve a thermally neutral situation whereby the heat generated by activity and metabolism (as affected by clothing insulation) is in balance with the heat loss due to the temperature differential of the room. PMV is able to undertake this prediction by being based on a model derived from actual large-scale data. PMV is thus an effective way of translating objective environmental measures into subjectively felt states, the corollary of which is to provide a more accurate assessment of human performance under different conditions.

It is not possible to suit everyone, however, so while the PMV index provides a good overall measure of 'sensation' the Predicted Percentage of Dissatisfied (PPD) measure provides a measure of 'dissatisfaction'. A graph (Figure 10.1) is used to plot the obtained PMV value, upon which it is possible to read across and obtain an estimate of the percentage of people likely to be thermally dissatisfied. Generally, the goal is to achieve a PPD value of less than 10 per cent with greater values normally taken as a cue to consider further intervention.

The following table (Table 10.3) has been created from the information and estimates contained in BS EN ISO 7730:1995 'Moderate thermal environments – Determination of the PMV and PPD indices and specification of the conditions for thermal comfort'. It is important to note that the military command and control environment is on the borderline between what is defined as a 'moderate' thermal environment and an 'extreme' one. In order to facilitate comparisons with alternative command and control domains and best practice BS EN ISO 7730:1995 will be used for the purposes of this analysis.

Table 10.3 shows a range of activity levels, air temperatures and their corresponding PMV and PPD values. The green shading shows the optimum temperature required to achieve a 'thermally neutral' condition, whilst the red shading shows the (approximate) actual temperatures achieved

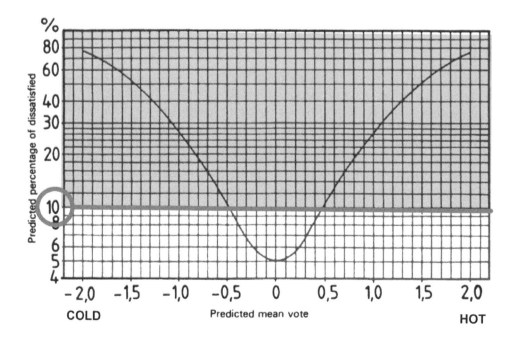

Figure 10.1   Graph showing how PMV values map on to the predicted percentage of people thermally dissatisfied. A value of 10% or less is desired

Source: BS ISO 7730: 1995, p. 6

Table 10.3   Summary table of control room relevant PMV and PPD values

Activity Level	Clothing	Air Temperature	PMV	PPD
SEDENTARY e.g. desk-based work undertaken whilst seated. Metabolic rate = approximately 70W/m2	WARM e.g. underwear with short sleeves, shirt, trousers, jacket, quilted outer jacket, army overalls	10	-0.77	19%
		12	-0.49	10%
		14	-0.21	6%
		16	0.08	5%
		18	0.37	8%
		20	0.67	11%
		22	0.97	26%
		24	1.27	40%
		26	n/a	n/a
		28	n/a	n/a
LIGHT ACTIVITY e.g. work undertaken whilst standing Approximately 90W/m2		10	-0.08	5%
		12	0.14	6%
		14	0.35	7%
		16	0.57	11%
		18	0.78	19%
		20	1.00	26%
		22	1.23	40%
		24	n/a	n/a
		26	n/a	n/a
		28	n/a	n/a
MEDIUM ACTIVITY e.g. work undertaken whilst moving and walking Approximately 116W/m2		10	0.34	7%
		12	0.70	18%
		14	1.07	26%
		16	1.45	50%
		18	n/a	n/a
		20	n/a	n/a
		22	n/a	n/a
		24	n/a	n/a
		26	n/a	n/a
		28	n/a	n/a

in both Bde and BG HQ, where this differs. In addition, the following modelling assumptions are made:

- A visual assessment of the clothing worn by personnel suggests that it falls within the 'warm' category and, therefore, a clo value (an estimated measure of the thermal insulation of clothes) of approximately 2.0 is a reasonable basis for calculating the PMV value.
- A similar visual assessment of physical activity was also undertaken. During the observation

period operators had varying levels of activity, ranging from seated at tables for prolonged periods working on digital MP/BM terminals through to standing whilst briefing, to greater degrees of movement and walking during other periods.

- Finally, note that BG and Bde HQ are partly comprised of Warrior armoured vehicles and these too are the subject of Environmental Analysis where appropriate.

Table 10.3 shows that 'on average' when personnel are engaged in light and medium activity, given the clothing they are wearing, they are in or around a thermally neutral situation even though the outright air temperature is far lower than recommended. However, the air temperature is still too cold in the context of more sedentary activity. It is important to note that digitisation in the form of digital Mission Planning and Battlespace Management (MP/BM) is, if anything, creating the conditions for precisely this sort of activity and, as such, creating a requirement for warmer HQ facilities.

The graphs that follow communicate the state of the thermal environment in more detail (Figure 10.2 and Figure 10.3). First of all it was noted that heaters were deployed in both HQs for the purposes of the current exercise, which may or may not be employed on live operations in similar conditions. In both cases the heating equipment was able to raise the temperature of the inside of the tent to around 5°C higher on average than the prevailing outside conditions (4.77°C for the Bde HQ and 5.68°C for BG HQ). This, along with the marked vertical temperature differential noted above in Table 10.2, suggests that the insulating properties of the tent are poor. In addition, there was noticeable water ingress due to damage and wear of the tent canvas which has a knock-on effect on relative humidity levels (see below).

In detail, the 19 spot measurements conducted at Bde HQ show that 18 of those intervals (95 per cent) measured lower than the 20°C minimum recommended by BS EN ISO 11064-6:2005. Even with the mitigating effects of appropriate clothing and non-sedentary activity, however, 12 intervals (63 per cent) were lower even than the 10°C thermally neutral condition recommended. In human performance terms this environment is too cold.

Much the same story is in evidence at BG HQ. The 16 spot measurements conducted here show that all of them (100 per cent) measured lower than the 20°C minimum recommended by BS EN ISO 11064-6:2005. Again, even with the mitigating effects of appropriate clothing and non-sedentary activity, 12 intervals (75 per cent) were lower even than the 10°C thermally neutral condition recommended, dipping below 5 °C on one occasion. In human performance terms this environment is also too cold.

Figure 10.4 and Figure 10.5 present the results of the spot measurements taken of relative humidity. In both Bde and BG, humidity levels regularly exceed the threshold defined by BS EN ISO 11064-6:2005, no doubt due to the fact that both HQs are constructed on bare ground and that the tents are not fully waterproof. At Bde HQ, 42 per cent of the measurements exceeded the threshold of 70 per cent whilst at BG HQ this was slightly higher at 44 per cent. Although not represented on the graph, BG HQ peaked at an alarming 92.5 per cent when located in a particularly harsh setting. Clearly this sort of level presents issues not just for human performance, but also potentially for the performance of digital battlefield systems and equipment, and by any measure is unacceptably high if good system performance is the goal.

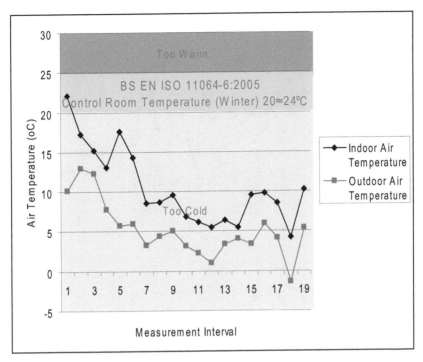

Figure 10.2   Longitudinal overview of the thermal environment extant in Bde HQ

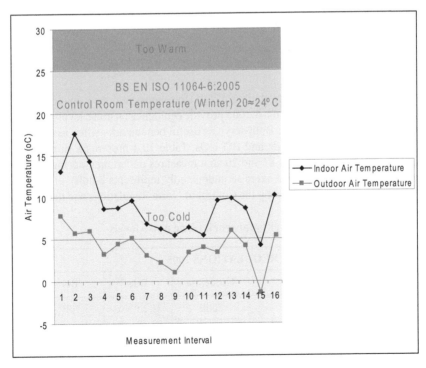

Figure 10.3   Longitudinal overview of the thermal environment extant in BG HQ

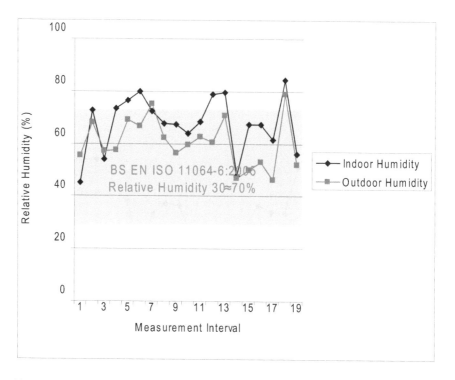

Figure 10.4 Longitudinal overview of relative humidity extant in Bde HQ

Acoustics

Minimum Requirements

The Control of Noise at Work Regulations (2005) provides guidance on noise levels that are potentially harmful. BS EN ISO 11064-6:2005 provides guidance on noise levels that are considered 'optimum' in control room settings. Both serve as useful benchmarks when assessing noise levels in the context of the exercise, at Bde and BG HQs. Table 10.4 presents a summary of what was found, based on averaging between 17 and 20 spot measures undertaken during the CPX.

Table 10.4 shows that whilst the exercise undoubtedly represents a noisy environment it is, on average at least, not of a level defined by the Control of Noise Regulations and the Health and Safety

Table 10.4 Minimum requirements for work place noise levels

CONTROL OF NOISE AT WORK REGULATIONS 2005				
	Guideline	Bde	BG	Warrior
Lower exposure action value	80dB	73.2dB(A)	73.7dB(A)	61.3dB(A)*
Upper exposure action value	85dB			*Warrior engine not running
Maximum daily or weekly exposure	87dB			

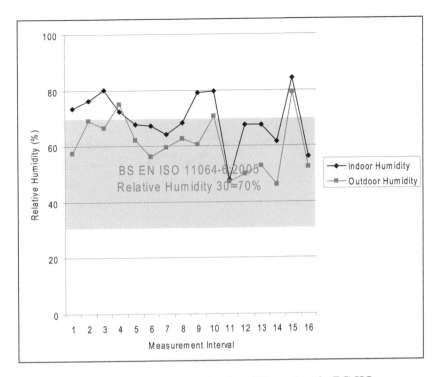

Figure 10.5   Longitudinal overview of relative humidity extant in BG HQ

Executive (HSE) as being harmful with prolonged exposure. However, as Table 10.5 shows, it is far from optimal either. Drawing on best practice in the civilian domain, it is recommended that similar command and control tasks are performed in considerably quieter conditions if human performance is to be optimised.

When the measures are analysed in detail it can be seen that noise levels, on occasion, reach quite high levels. Figure 10.6 shows the results in detail for Bde HQ and the chart shows that for a period the noise levels become very close to the HSE's lower action value of 80dB. This is the point beyond which remedial measures should be taken to protect personnel. The chart further illustrates the fact that noise levels, whilst generally not at harmful levels, are in a region that is considerably above optimum.

Table 10.5   Optimum requirements for control room noise levels

EN ISO 11064-6: 2005				
	Guideline	**Bde**	**BG**	**Warrior***
Maximum ambient noise	45dB LAeq.y	80.3dB(A)	80.3dB(A)	61.3dB(A)
Maximum background noise	35 dB LAeq.y	82.5dB(A)	82.5dB(A)	82.5dB(A)
Minimum background noise*	30 dB LAeq.y	67.8dB(A)	70.2dB(A)	69 dB(A)

*Warrior engine not running onto dB LAeq.y

Note: dB(A) maps approximately

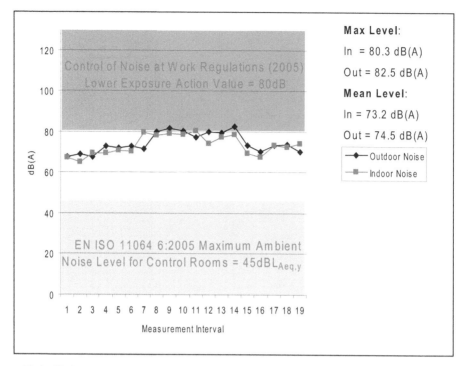

Figure 10.6   Noise levels measured in dB(A) at Bde HQ during the CPX

The same picture emerges for BG HQ (see Figure 10.7). Noise levels here also enter into the HSE's lower exposure action value range on a number of occasions, whilst being generally well above optimum levels recommended by BS EN ISO 11064-6:2005.

Human Factors Best Practice

In terms of best practice a good strategy to employ, one that requires only basic equipment but goes beyond minimum requirements, is the 'rapid sound quality assessment' technique (Torres, 2005). This method comprises three steps and the use of a sound level meter with A and C weighting curves (a common feature). The three steps are as follows:

1. Configure the sound level meter to read in dB(A) and take a number of measurements at individual workstations. Take the average of the readings at each workstation.
2. Configure the sound level meter to read in dB(C) and take a number of measurements at individual workstations. Take the average of the readings at each workstation.
3. Subtract the averaged dB(C) reading from the averaged dB(A) reading.

The difference value contains useful information concerning the spectral character of the ambient sound in the control room, where:

- 10dB or greater relates to the ambient noise having a 'hissy' character. These mid to higher frequency components can potentially affect the intelligibility of dialogue and may even have a subtle masking effect on alarm sounds in more extreme cases.
- ≈ 15 dB is subjectively neutral, the ambient noise being biased in neither high nor low

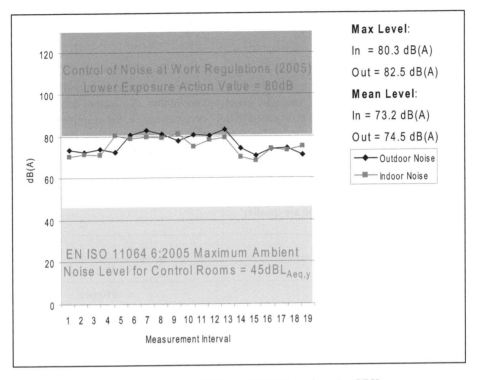

Figure 10.7   Noise levels measured in dB(A) at BG HQ during the CPX

frequency directions.

- 20 db or higher suggests that the ambient sound has a low frequency rumbling character. This is a clue that excessive low frequency content may be present (Torres, 2005) which, in itself, has been associated with poor cognitive performance. In terms of auditory perception, low frequency sound is omni-directional so subjectively it can be difficult to take action in order to alleviate the sensation (that is, moving around the room or even localising the source is difficult and/or has no effect).

The results in Table 10.6 show that the spectral content of the ambient sound is biased somewhat in favour of higher frequency components in all locations, but particularly in the Bde HQ. Here it was noted that the use of multiple portable generators outside of the tent seemed to be the primary cause. The intelligibility of speech and the discrimination of any auditory alerts or warnings can be affected by ambient sound of this character.

Table 10.6   Rapid assessment of spectral character of ambient sound

	Guideline	Bde	BG	Warrior
dB(C) – dB(A)	15dB = neutral/acceptable	9.8dB	10.4dB	12.3dB
		Balance in favour of higher frequencies		

Air Quality

Minimum requirements

The overriding air quality requirement is that 'the control room should be supplied with outdoor air in sufficient quantities to dilute internally generated pollutants'. Air quality does not have quite such a fixed subjective experience (compared to light or temperature, for example) but air quality issues can result in smells and odours and in a general feeling of 'stuffiness'. Problems with air quality have also been highlighted as giving rise to longer-term diffuse problems such a deleterious feelings of malaise and lethargy. The objective values and thresholds below which such sensations can hopefully be avoided are shown in Table 10.7.

Table 10.7   Objective values for air quality

EN ISO 11064-6: 2005	
Per Person Fresh Air Supply	29m3/h
Maximum CO2 concentration	910ppm

In the case of the exercise it was observed that there was little in the way of active ventilation in place in order to provide a $29m^3/h$ fresh air supply; on the contrary in fact, there was very little active or passive ventilation. Whilst at Bde HQ it was noted that hot air blowers were in use, at BG HQ, however, oil/paraffin heaters were employed inside. At BG the command tents were all located on a covered hard standing meaning that when the engines of the Warriors were running, exhaust fumes collected quite markedly under the roof and were perceptible, by odour at least, within the tents on occasions. Unfortunately carbon dioxide/monoxide measurements require specialist test and metering equipment that are beyond the scope of this rapid assessment. However, several techniques were employed to assess the subjectively felt air quality experience of personnel in both locations.

Walk-Through Analysis

The first analysis takes the form of a basic walk-through, checking the environment(s) for compliance with features and guidelines stated in BS EN ISO11064-6:2005. In the current analysis the main tents of Bde and BG HQ were assessed, followed by the Warrior armoured vehicles attached to both (see Table 10.8).

The quick walk-through highlighted several areas of 'non-compliance' with normal control room standards. In so far as both main tents go, the only area where they are compliant with standards is in the provision of protective equipment in the form of Nuclear, Biological and Chemical (NBC) suits. The Warriors, by their enclosed and rugged nature, offered a higher degree of protection and had onboard filtered ventilation and heating systems.

Environment Survey

Having performed the initial walk-through assessment, the analysis proceeded with a standard environmental survey instrument (Hedge & Erickson, 1998) that polled individual control operators on their subjective experience of air quality, as shown in Figure 10.8.

Table 10.8 Results of air-quality walk-through checklist

	Bde Tent	BG Tent	Warrior
Draughts?	Yes	Yes	No
Intakes too close to outlets?	n/a	n/a	No
Noise?	Yes	Yes	Yes
Filtered air supply?	No	No	Yes
External air pollution sources?	Yes	Yes	Yes
Prevent malicious introduction of material?	No	No	Yes
Protective equipment available?	Yes	Yes	Yes

The results of the questionnaire are presented in two parts: environmental conditions and symptoms. It was concerning that in both cases respondents reported conditions and symptoms in all categories (see Figure 10.9).

The overriding environmental condition that respondents noted was the fact that it was too cold. 100 per cent of BG respondents thought this as did 85 per cent of Bde (this of course serves as further evidence to support the above section on ambient temperature). Next in line was the rating for odour. Approximately 75 per cent of BG respondents felt this to be unpleasant as did 71 per cent at Bde. In addition, it was felt that the Bde tent was too dusty and too dry (57 per cent of respondents thought this). This may be an artefact of the hot air blowers perhaps, as humidity was, in actual fact, a little on the high side.

All respondents, whether located at Bde or BG HQ, reported minor physical symptoms attributable to the conditions they were working in (see Figure 10.10). Stuffy nose (86 per cent), tiredness (71 per cent) and mental fatigue (55 per cent) were noted for the Bde HQ, tying in, perhaps, with the data on the air being too dry and dusty. Despite an ostensibly harsher environment the BG personnel didn't report individual symptoms as being particularly prevalent. What is in evidence, however, is a solid showing for most of the symptoms, headaches aside. Both these and the previous findings reveal the environmental effects on personnel, and whilst only based on a small sample and simplistic measurement approach, represent evidence that this issue warrants further exploration.

Lighting

Minimum Requirements

Lighting is provided in the Bde tents by fluorescent tube lighting, whilst in BG it is provided by halogen lights. Inside the Warrior vehicles lighting is provided by miniature fluorescent tubes in bespoke casings. There are no windows or openings that admit daylight directly into the tent. The canvas material is also matt green in colour and highly light absorbent.

BS EN ISO 110640-6:2005 states that control rooms should be lit to a maximum level of 500 lux if VDUs are in use, 750 lux if not. Minimum requirements for control room lighting are shown in Table 10.9, compared to the values achieved during measurements conducted in both locations.

Environment Survey

Please answer the following questions about environment comfort condition and health symptoms that you may have experienced in the office during the past month (4 weeks).

What is your gender? Male ☐ Female ☐

Please indicate whether you have experienced any of the following environmental conditions in the workplace over the past month:

	Yes	No
Air temperature too cold	☐	☐
Air temperature too warm	☐	☐
Too little air movement	☐	☐
Air too dry	☐	☐
Unpleasant odour in air	☐	☐
Air too stale	☐	☐
Air too dusty	☐	☐

Please indicate whether you have experienced any of the following symptoms at least once over the past month and whether they got better when you were away from the office (e.g. in the evenings or at the weekend):

	Yes	No
Irritated, sore eyes	☐	☐
Sore, irritated throat	☐	☐
Hoarseness	☐	☐
Stuffy, congested nose	☐	☐
Excessive mental fatigue	☐	☐
Headache across forehead	☐	☐
Unusual tiredness, lethargy	☐	☐

Figure 10.8   The Cornell Office Environment Survey

Source: Hedge, 2005, pp. 64–9

Based on these results it is possible to conclude that the land warfare command environment is considerably darker than control room standards recommend. This has clear disadvantages for human performance, in particular with the advent of digitisation. For example, the legibility of text can become an issue as can the visual accommodation required to switch between the UDT's bright backlit screen and analogue materials located away from it.

Human Factors Best Practice

An office lighting survey was deployed to assess the adequacy or otherwise of the lighting provided. Table 10.10 presents the results of the survey, which gathers subjective impressions from

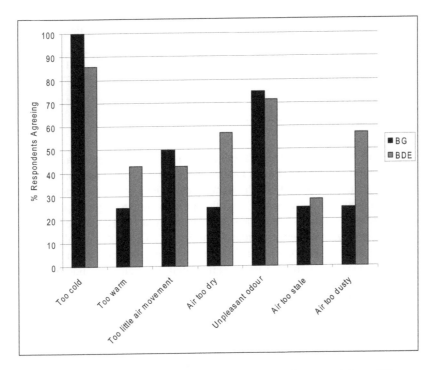

Figure 10.9   BG and Bde responses to questions about environmental conditions

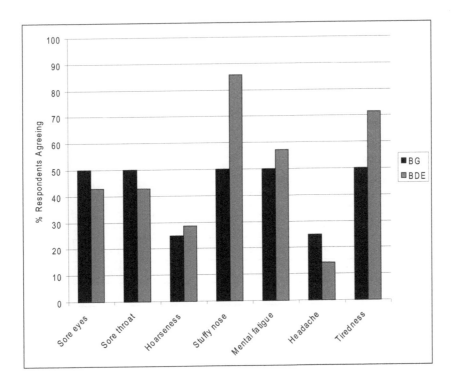

Figure 10.10 BG and Bde responses to questions about physical symptoms

Table 10.9   Maximum and minimum requirements for control room lighting

EN ISO 11064-6 2005	Guideline	BDE and BG Tent
Maximum illuminance levels	500lx	97lx
Minimum illuminance levels	200lx	48.5lx

respondents and maps them on to the results gained from a much larger population, thus providing a useful reference point. This rapid Human Factors tool allows a useful assessment to be made.

The difference of subjective rating of the lighting conditions from the norm is revealed above by the dark shading (showing a divergence from the norm to an extent greater than 10 per cent). It is clear that the lighting conditions inside the smaller halogen-lit BG HQ are felt to be subjectively poorer, relatively speaking, than the larger fluorescent-lit Bde HQ, but it is not all bad news. Despite this, a greater proportion than the norm in both locations felt the lighting conditions to be comfortable. Overall, though, it appears that specific issues surrounding the lighting conditions in BG HQ are

Table 10.10   Office lighting survey (Eklund & Boyce, 1996)

SURVEY QUESTION		PER CENT AGREEING WITH STATEMENT		
		Population	Brigade	Battle Group
Overall, the lighting is comfortable.		69	83%	100%
The lighting is uncomfortably bright for the tasks I perform.		16	0%	0%
The lighting is uncomfortably dim for the tasks I perform.		14	17%	0%
The lighting is poorly distributed here.		25	0%	75%
The lighting causes deep shadows.		15	17%	75%
Reflections from the light fixtures hinder my work.		19	33%	50%
The light fixtures are too bright.		14	17%	0%
My skin is an unnatural tone under the lighting.		9	0%	25%
The lights flicker throughout the day.		4	17%	17%
How does the lighting compare to similar workplaces in other buildings?	Worse	19	0%	75%
	About the same	60	100%	25%
	Better	22	0%	0%
Able to read 8 point print and larger.		99	100%	100%
Able to read 6 point print and larger.		94	100%	75%
Able to read 4 point print and larger.		76	67%	50%

* Dark shading denotes a difference < +/- 10 per cent

subjectively felt to be poorer, implicating the halogen lighting in relation to reflections hindering work, being poorly distributed, comparing worse against similar workplaces and giving skin tone an unnatural colour. This last factor relates to the Colour Rendering Index (CRI).

BS EN ISO 11064-6:2005 states that lighting should posses a CRI of over 80. The perceived colour of illuminated objects depends on the light that is being shone on them. Sodium street lighting, for example, is termed monochromatic because everything will appear as varying shades of orange; as such, its colour rendition is very poor (a CRI of zero would be typical of this light source). An incandescent bulb, on the other hand, facilitates excellent colour rendition and has a CRI of 100. Tube lighting of the sort used in the Bde tent normally has a CRI ranging from approximately 60 to 90. It was not possible to verify the specific make and model of tube used in either the Bde main tent or the Warrior vehicles, but triphosphor varieties are generally favoured for their high CRI values and in an environment adapting to the requirements of UDT displays and Smartboards these should be considered if not already fitted. Whilst on the topic of fluorescent lighting, high frequency systems are also recommended to reduce flicker effects; again, it was not possible to assess this characteristic during the present exercise.

In the BG HQ the tungsten halogen lights in use typically have CRI ratings in the region of 95. The issue here seems to be with their inappropriate directivity (and resultant glare) more than it is with colour rendering. A switch to high frequency, high CRI fluorescent tubes might be appropriate.

Vibration

BS EN ISO 11064-6:2005 does not specify minimum criteria for objectively measurable vibration. Many standards do indeed refer to vibration but this is typically in relation to whole/part body vibration of the sort caused by operating machinery or vehicles; generally speaking it is not relevant to control room design. In terms of minimum requirements compliance revolves around:

- Positioning the control room the maximum distance possible from vibration generating sources (such as generators and compressors).
- Insulating the control room from vibratory sources using, if necessary, vibration absorbers in the floor, ceilings and walls.

Human Factors best practice

Bde and BG HQs were both located directly on firm ground. There were numerous noise and vibration generating elements nearby but the lack of any 'structure' (in the normal sense) meant that such vibrations were not sufficient to impart those sensations directly into the ground and thus affect the command tents. A combination of walk-through checklists and questionnaires completed by personnel themselves revealed no evidence to suggest a problem with vibration.

In addition to the above, reference to the section above on acoustical measurements, in particular the rapid sound quality assessment, provides a quick assessment route into the presence of problematic structure or ground borne vibration. If these initial checks reveal that such stimuli exist in the control room, then further investigation to identify the source is required along with accompanying intervention. This sort of specialist intervention is beyond the scope of the current book and thus the analysis need not be pursued further.

Interior Design and Aesthetics

The title of this section seems incongruous in a military setting but it needs to be pointed out that as a general term, referring to background colour, materials and so on, it has much to say in terms of human performance with digital systems.

Minimum requirements

Minimum aesthetic and interior design requirements can be quickly assessed in the context of a walk-through analysis covering the following items:

- a calming backdrop to control activities;
- excessive use of dark or light finishes is avoided;
- pale, untextured wall finishes are in use;
- excessively strong patterns are avoided;
- heavy contract grade carpets with small random pattern are used;
- ceilings lighter than walls, walls lighter than floors;
- humanisation of the environment (for example, plants, furnishings and so on) is present;
- in general, the selection of finishes, furnishing and design is suitable for 24-hour operation.

The results of this analysis for the exercise are shown in Table 10.11.

It seems a little trite to assess the military context too harshly in this manner, so it is important to point out that what is not being recommended is the sort of softly furnished control room found in civilian domains. On the other hand, and recognising fully the constraints of the military environment, the results do suggest that there is a lot of scope for improvement and enhancement. For example, instead of 'carpet' perhaps matting or flooring could be used in some situations? Perhaps a different control room layout might facilitate a more calming setting for safety critical work? The aim of this analysis is to raise these issues in a structured and systematic manner on the premise that they have human performance implications.

Under the heading of interior design and aesthetics, BS EN ISO 11064-6:2005 is quite specific in terms of reflectance values. For the purposes of the standard, reflectance is defined as the 'ratio of the luminous flux reflected from a surface [...] to the luminous flux incident on it'. Note that reflectance also depends on the direction from which light strikes an object and its spectral content. For practical purposes if all the incident light is reflected off of a surface then the reflectance value (ρ) would be 1 and the surface would be something akin to a mirror. Somewhat less reflectivity is specified in the standard, as shown in Table 10.12 where these guidelines are compared to the values actually achieved.

Reinforcing some of the observations made above in the section on lighting conditions, not only are light levels low but reflectance values are too. In basic terms, as one would expect, the canvas and other materials used in both HQs absorbs a considerable amount of light, more than is recommended by BS EN ISO 11064-6:2005. This again is not to recommend having a colour scheme identical to a civilian control room but to highlight areas for improvement and opportunities to better support digitisation.

Human Factors best practice

Various procedures for checking glare can be easily adopted for other surfaces to provide a quick check requiring the minimum of equipment. The analysts can survey walls, floors and other

Table 10.11   Results table for interior design and aesthetic checklist

	Bde	**BG**	**Warrior**
A calming backdrop to control activities	No	No	No
Excessive use of dark or light finishes is avoided	No	No	No
Pale, untextured wall finishes are in use	No	No	No
Excessively strong patterns are avoided	No	No	No
Heavy contract grade carpets with small random pattern are used	No	No	No
Ceilings lighter than walls, walls lighter than floors	No	No	Yes
Humanisation of the environment (e.g. plants, furnishings etc.) are present	No	No	No
In general, the selection of finishes, furnishing and design is suitable for 24-hour operation	No	No	No

Table 10.12   Control room aesthetics and surface reflectivity

BS EN ISO 11064-6:2005			
Control room element	Reflectance value (ρ)	Bde	BG
Floor finishes	0.2 – 0.3	0.01	0.02
Wall finishes	0.5 – 0.6	0.03	0.13
Partitions etc	0.5 – 0.6	n/a	n/a
Ceiling	< 0.8	0.04	0.19

*An average value of 72.8lx is used for incident light

potentially specular surfaces for glare and high reflectance. The fact that it can be readily noted by the naked eye suggests the presence of reflectance above $\rho = 0.3/0.6$. In these cases a mirror can be placed over the top of the noted glare point and the source of the glare identified. Appropriate intervention can then take place.

Note also that the Office Environment Survey dealt with above (Figure 10.8) also provides valuable information/prompting in terms of interior design and aesthetic problems.

Summary

The opportunity was taken to adopt a systems approach to digital MP/BM and to examine the wider environmental factors that it resides within. This was a brief rather than full environmental assessment, yet some major issues were highlighted when the exercise context was compared to acknowledged best practice in the civilian domain.

The overall analysis granted the exercise working environment an opportunity to comply directly with 42 items. As Figure 10.11 shows, Bde HQ failed to comply with 35 of those whilst BG failed to comply with 40.

In summary the results of the analysis reveal that:

- The command and control environment is generally too cold, despite warm clothing being worn. To some extent this is a reflection of the increased amount of sedentary activity that digitisation requires.
- Noise levels approached and sometimes exceeded the HSE's lower exposure action level of 80dB, indicating the presence of harmful long-term levels. Noise levels are considerably in excess of what standards would regard as optimum given the nature of the task being performed. The spectral content of the noise is also concerning as currently it presents a possible risk in terms of voice and alarm intelligibility.
- Air quality also rated very poorly in terms of facilities (or lack thereof) to support a clean fresh air supply. A considerable number of respondents reported unpleasant odour and all respondents noted physical symptoms that they attributed to the conditions (including tiredness and upper respiratory tract ailments).
- Lighting conditions were measured and failed to meet relevant standards. The control room setting is too dark and in the case of BG in particular the lighting has poor directivity and excessive glare.

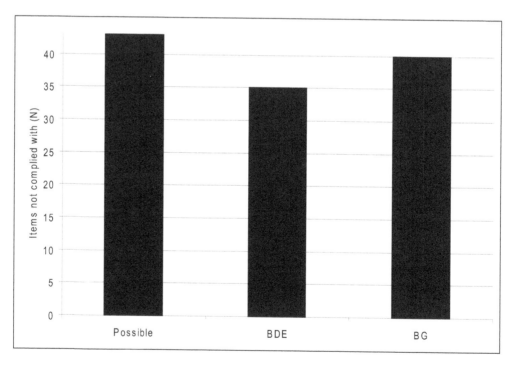

Figure 10.11 Bar chart showing the extent of non-compliance with environmental guidelines

- Despite numerous vibration generating elements in close proximity to both command centres, vibration did not emerge as a significant problem.
- Finally, the human performance aspects of interior design, materials and aesthetics in Bde and BG main tents failed every comparison with relevant guidelines. Excessive darkness, lack of adequate flooring (in BG), and lack of fitness for purpose in terms of 24-hour operation were in evidence.

To conclude, the comparison undertaken between best practice and guidelines extant in the civilian domain and those extant in the exercise context is not to suggest that the military setting is directly comparable to its civilian counterparts nor should it emulate it completely. With the advent of digitisation, though, environmental requirements change and the present comparison is designed to raise awareness of this and the issues that have to be considered. Regardless of all this, by any measure the environmental conditions are certainly harsh for people working in command and control HQs when compared to their civilian counterparts. These conditions are likely to place additional strain on the people and technology in performance of their tasks.

Chapter 11
Summary, Conclusions and Recommendations

Summary of Findings

The aim of this book has been to assess a digital Mission Planning/Battle Management (MP/BM) system in an operational field trial, and where possible to compare it with the analogue process that came before it. Significant shortcomings in the system's performance were identified which arise not necessarily because the system is technically ineffective (although this is sometimes the case) but principally because it is not well matched to its human users. In other words, the interface between the capabilities provided by the system and the ability for human users to access them, and harness their benefits, is in some cases critically hampered. The findings of the book, therefore, provide a salutary lesson in the dis-benefits that arise from not integrating Human Factors (under the rubric of Human Factors Integration) right at the beginning, and then throughout, the complete design life cycle (as advocated in chapter two). What might at first appear to be an ostensibly technical system is in fact a 'socio-technical system'. Human Factors provides the tools and techniques to optimise both.

On a more practical level, what the findings of this book reveal is that the introduction of a simple and naturalistic facility such as encrypted radio has yielded a significantly positive effect on the complex tasks involved in tactical command and control. Analysis within this book finds that the radio is the main conduit for this improvement over the analogue process. Simple 'Chat' and 'Free Text' tools appeared to offer benefits far beyond those anticipated, and were preferred to the more 'formal' tools that had required significant design and development effort. Whilst the highly complex digital MP/BM system improved activity in some areas, it significantly hampered it in others. Numerous shortcomings in nearly all aspects of what Human Factors is normally considered to be about, that is, the Human Computer Interface, were encountered. These include:

- the keyboard and thumb cursor (which were both difficult to use so external keyboards and mice were used instead);
- screen (one screen was not enough – users would need two as a minimum to do their work);
- interface design (the interface was not intuitive and did not conform to the generally held software conventions);
- processing power (there is a mismatch between the available processing power and the requirements of the software);
- logging on (logging on to the system is unnecessarily complex);
- contact reports (locations of enemy engagements) and friendly forces position reporting lag far behind the true situation.

An analysis of the planning process found that workstations really only supported two parts effectively, namely the editing of the Warning Order (WO) (after a paper printout had been reviewed and relevant sections highlighted) and writing up of the Requests For Information (RFI) for higher command (after the Mission Analysis had been committed to a whiteboard). In both of these instances, the digital MP/BM system was used after the conventional media had been used to extract meaning. Thus digital MP/BM was being used as a conduit for passing requests and information on to others, rather than as a tool to aid the intellectual activity and creativity inherent in the planning process which was observed when looking at the pre-digitised system (chapter three). The key finding here is that the 'process' of planning is as important, if not more so, than the eventual 'products' of planning.

Based upon the evidence collected it is not possible to make a judgement about the robustness of the data and voice services. The analysts did however observe numerous system crashes and extremely slow data transfer times. There is little evidence to suggest that the new system supported the staff process or increased tempo. On the contrary, subjective opinion (see chapter four) as well as operation timings indicates that the digital MP/BM system has had a negative impact on tempo and offers little above the analogue process. The results of the survey discussed in the constraints analysis (chapter four) revealed that whilst the system was rated positively for its ability to support battlefield management, it scored poorly for its ability to support planning, with none of the respondents offering a positive opinion. More specifically the summary of opinion shows that the new system was scored as worse for: tempo, flexibility, efficiency, effectiveness and fidelity as well as the time taken to produce plans. The introduction of the encrypted radio is clearly a very positive and welcome addition, and examination shows this attribute for the majority of the positive opinion of battlefield management in the digital system. Examination of the staff officers' ratings of aspects of the workstations, however, reveals the software and digital planning tools are incompatible with the planning activity. In some cases, the digital MP/BM software could be improved through redesigning the user interface and interaction experience, however, in many cases the planning activity should not be attempted in this digital system.

The analysis of digital MP/BM using Hierarchical Task Analysis (HTA) and Systematic Human Error Reduction and Prediction Approach (SHERPA) (chapter five) showed significant shortcomings in the structure of tasks and that the system had many opportunities for users to commit errors. The most significant shortcomings of the system were summarised in the following points:

- lack of standard conventions (such as the copy and paste process where there is already an industry standard approach);
- overly complex processes (such as marking up the maps and development of the overlays);
- complex menus (with structures that are too wide and deep);
- unintuitive drawing tools;
- lack of consistency in the user interface;
- general poor interface design.

Some of these points were picked up by other analyses, such as observation, the usability questionnaire and the Engineering Equipment & Materials Users Association (EEMUA) survey, which indicated that they were of major concern.

Analysis of the Distributed Situation Awareness during the planning phases (chapter six) revealed questions about the timeliness and accuracy of information, the tempo of planning in digital MP/BM, the accuracy of information from digital MP/BM (as staff had to engage in additional checking activities) and the poor support for the different Situation Awareness (SA) requirements in the different planning cells. Analysis of the operational execution tasks revealed that the Local Operational Picture (LOP) was often out-of-date or spurious (clarification of Own Situation Position Report (OSPR) data was required, placing more load on the secure voice channel for updates of the blue force positions) and a low level of LOP and OSPRs (requiring the operations cell to compensate for digital MP/BM's shortcomings by drawing blue force positions directly on to the Smartboard – but these were wiped off every time digital MP/BM updated or was changed). In summary, it was found that Distributed Situation Awareness was not well supported by the digital MP/BM system, as people have different (SA) requirements, which are subject to change, depending upon their tasks and goals at any moment in time.

The Social Network Analysis (SNA) (chapter seven) provided a unique insight into the real-life challenges of introducing net-centric platforms in service. Emergent behaviours arose in which 'something stupid', in this case a highly simplistic Free Text facility, 'bought something smart'. Whether this was an attempt by users to restore the mismatch between their net-centric approach and their corresponding cold-war style problem is debatable, but in either case it tells systems designers that users favoured a 'simple system that enabled them to do complex things' rather than a 'complex system that only enabled them to do simple things'. Another interesting issue was that the presence of data, or even the network to carry it, was not enough on its own to increase operational effectiveness. The ability to turn that data into information or knowledge through human actors ostensibly 'doing something with it' is key. The results of the present analysis suggest that not enough of this is occurring in the present system and that greater attention toward the user/ technology interface, and perhaps even the entire philosophy behind digitisation, requires further evolution aided by the body of knowledge contained in Human Factors.

Another finding that goes to the heart of the NATO model of command and control (also chapter seven) is where the live organisation actually positioned itself in practice. Its static characterisation clearly bore the hallmarks of a distinctly net-enabled centre of gravity, as defined by procedures and doctrine, but it was the users, in attempting to meet the challenges created by their 'problem', that pulled this organisation into virtually all areas of the approach space. By and large it was down to them and their interpretation of the system, the way they massaged it and made such adjustments as they saw fit, that gave the system its resultant levels of agility and tempo. The fact that the system, in the end, was able to provide such a level of agility and tempo is encouraging. But as mentioned above, there was a fundamental mismatch between approach and problem which seemed very hard to overcome. Indeed, success in the mission required arduous efforts on the part of those involved and to some extent occurred despite the presence of the MP/BP system rather than because of it.

From the 35 EEMUA 201 principles against which digital MP/BM system was evaluated (chapter eight), only eight were met (where no improvements were required, that is, window resizing, single and multiple windows, snap shot capture, power back-up, pictorial displays, cool background and foreground colours, no use of animation). A further 12 principles were partially met (where some improvements to the current system are recommended, that is, permanently viewable overview display, capability to expand screens, provision of start-up, state monitoring and shut-down screens, ease of navigation through hierarchy, simple hierarchical structure, ease of navigation across levels, user-defined default configurations, user-defined favourites and quick access buttons, electronic shift handover notes, quick and unambiguous item selection, different interaction methods, sparing use of colour coding), whilst eight principles failed to be met (where significant shortcomings in design were identified, that is, access to necessary information, maximum number of windows on display, important windows on permanent display, hotlinks to databases, quick and consistent feedback, quick access to details, uncluttered displays and single task-based windows). A further seven of the EEMUA 201 principles were deemed not applicable to digital MP/BM.

Further usability assessment was undertaken with a Human Computer Interaction (HCI) questionnaire (chapter nine) which supports the observational studies and EEMUA 201. It too shows many shortcomings in the interface design, as rated by those who were charged with using it. It is interesting to note that overall ratings were generally lower at Battle Group (BG) level (where the system is being used very much 'in the field'), but even at Brigade (Bde) (where the environment within which the system is being used is a little more benign) the overall ratings failed to go beyond neutral. The system was rated particularly low on 'explicitness' and 'error prevention and correction'. Generally speaking, the people using the system did not find it intuitive and many found that they lost work due to inadvertent errors.

On the topic of the environment within which the MP/BM system is designed to be used, for the sake of completeness it too was assessed (in chapter ten). This was not intended to inform the design of the digital MP/BM per se, rather it was to consider if the surrounding environment met with current standards for control centres (that is, BS/EN/ISO 11064-6:2005 Environmental Requirements for Control Centres). Seven main environmental criteria were evaluated: ambient temperature, humidity, air quality, lighting, acoustics, vibration and aesthetics. Some points of concern were highlighted. Although it should be noted that these criteria were developed for civilian rather than military control centres, environmental stressors are not trivial. Coupled with the fact that staff can work for shifts of 12 hours or more, it places even greater emphasis on the design of interfaces to match expectations, enhance ease of use and be error tolerant.

Recommended Improvements for Digital MP/BM

The extensive analyses undertaken in the course of this book has led to the identification of many potential improvements to digital MP/BM systems. Generally, the system is too slow, cumbersome, overly complex and does not use accepted conventions in the user interface. The main recommendations vary between short- and medium-term improvements. The improvements are divided into five sections: general design, user interface, hardware, infrastructure and support. General design improvements:

- Only digitise what needs to be transferred to other parties. The analogue system already works well for most of the planning process.
- Adopt a multimedia approach, such as the use of digital capture of images and text. Products that can be produced quickly using the analogue media may then be transferred into the digital medium.
- Take care to ensure that the digitised processes are as least as quick and convenient as the analogue processes, and that progress is not hindered by using the digital medium. For example, the digital Combat Estimate seven questions planning process might have to be undertaken quickly, for example, allowing only 17 minutes or less per question.
- The start-up processes need to be automated.

User interface design improvements:

- The Graphical User Interface (GUI) needs to adopt accepted conventions, such as those used by Microsoft, IBM and Apple, which are in the open literature. These conventions are familiar to the Headquarters (HQ) staff.
- The menu and menu navigation structure needs to be simplified.
- Provide a 'See All' button to allow a HQ to see everything within its boundaries.
- Excessive and additional functionality needs to be hidden from the user.
- Standard drawing tools need to be provided.
- Icons need to be grouped by function, to prevent accidental activation (for example, sending a contact report, rather than publishing a document).

- Consistency is needed in the interaction, dialogue and actions required by the user.
- Printing needs to be simplified, by the provision of a printing wizard, a print preview and a general reduction in the number of steps.
- Printing time needs to be reduced considerably.
- The interface needs to be customised to the role of the staff, for example, different interfaces are required for plans, operations, intelligence, fire support, engineering, I-Hub, logistics, watch keeper, air/aviation, administration and communications.
- Improve the refresh rates for pan and zoom (with mapping on) as these are far too slow.
- Limit the maximum number of windows that can be displayed at any one time.
- Reduce the requirement for users to have multiple windows open to perform a task, by designing task-based windows that contain all the necessary information to perform that task.
- Allow users to save 'favourites', for example, a palette of tools.
- Provide hyperlinks to electronic Standard Operating Procedures (SOPs).
- Provide immediate and relevant feedback on user actions.
- Reduce the amount of clutter on screens.
- Clear all forms as a default option.
- Provide meaningful error messages with suggested user actions.

Hardware improvements:

- The User Terminals are far too slow and faster processors are clearly required to support the current level of functionality. Consider the option of TEMPEST protection for the whole HQ tent, therefore allowing staff to use Commercial-off-the-Shelf (COTS) technology such as 'Tough-books' (for example), which have a more acceptable processing speed and considerably reduced cost.
- Increase the size and resolution of the screens, so that mapping visualisation will match HQ requirements.
- Most users need two screens, one for the current LOP and one for the work they are undertaking. The Ops cell needs three screens, one for the LOP, one for the staff work and one for the flash messages.
- Overlays need to be printed on A0 plotters. At present BG was required to print off the overlay on an A4 paper printer and redraw on to a map overlay.

Infrastructure improvements:

- Updating of the blue and red picture needs to be in real-time. Because of delays in the LOP update, the staff rely even more heavily on voice for updates thus exacerbating the delays for data. Greater bandwidth (or optimisation) is required to support both the levels of secure voice transmission observed in the exercise and rapid transmission of positional report data. Otherwise considerable reductions in secure voice transmission will be required.

Support improvements:

- Provide a simple user manual, with a 'Getting Started' section covering the basics for each cell. Examples of 'Getting Started' sections can be found in many COTS software manuals.

- Provide a help section that provides instructions on how to use an item as well as a description of the item (currently only a description is provided). For example, the current help section describes how to open a new user-defined overlay but it doesn't instruct the user on how to draw on the overlay.

Principles for Future Systems Design

The analysis presented here identified a series of problems associated with the design of the digital MP/BM system; the aim of this section is to introduce some approaches that can be used to aid system design. Of course, we readily acknowledge that many of the topics to be discussed may be uneconomical to implement into the current version, however, they should be of benefit for future programmes.

As alluded to already, the digital MP/BM programme has served to illustrate the implications of failing to consider and act upon basic Human Factors and interaction design principles in the early stages of, and throughout, the design process. In the early stages, input can be included at minimal cost resulting in massive cost saving and enhanced capabilities in the latter stages of the programme. These cost savings can be realised in the areas of reduced training cost and reduction in design changes.

Figure 11.1 shows the four key factors that can be manipulated to affect system performance. It is possible to map these four overlapping factors on to a set of four themes that emerged throughout the analysis.

Equipment vs. Technology

Many requirements-led procurement processes, ironically, lack the agility, tempo and self-synchronisation of the system trying to be procured. Changes are difficult to incorporate as requirements become fixed to contracted deliverables. Equipment specification is likewise fixed for long periods of time while technology continues to progress, thus bespoke systems with closed architectures tend to be the norm rather than COTS systems with more open architectures. The type of equipment best able to meet the aims and aspirations of NEC is likely to be an open, flexible sort of technology. This philosophy enables users to readily adapt it to suit their needs and preferences, and to genuinely support through-life capability as users evolve the technology, and technology is evolved to users.

'Train it out'

In the introduction to the text titled 'The design of future things', Norman (2007a) is keen to defend the users of systems. Norman makes the following comment, which seems pertinent to this case:

'The 'blame-and-train' philosophy always makes the blamer … feel good: if people make errors, punish them. But it doesn't solve the underlying problem. Poor design and often poor procedures, poor infrastructure, and poor operating practice, are the true culprits: people are simply the last step in this complex process… We must design our technologies for the way people actually behave, not the way we would like them to behave.'

(Norman, 2007a)

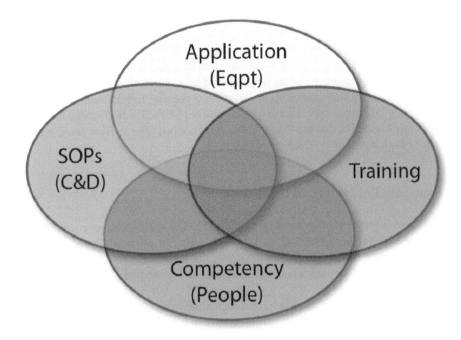

Figure 11.1   Key enablers to enhance performance (adapted from Macey 2007b)

Perhaps the most commonly adopted approach to system improvement within the UK Army is training, and the culprit when things do not work as planned is a lack thereof. There is undoubtedly benefit to system performance on more intensive and expensive training programmes, and it is to the credit of military personnel that in most cases they can (often with significant effort) overcome such challenges through training alone. Training is, however, no substitute for effective systems design. It is impossible to completely train for error prevention. Skill fade in unintuitive complex systems is a significant concern. Under times of duress and stress, it is highly likely errors will be made as people revert to stereotypical conventions and expectations when interacting with systems (Reason, 1997). In other words, it is not possible to completely 'train out' problems.

The Playstation generation

Training is one method to increase staff competency. The other approach is to recruit staff with a different, more compatible, skill and ability set. The problem, of course, is that the specific skill set required to operate an overly complex MP/BM system is not (or should not) be the focus of otherwise skilled and adept military personnel. In other words, they know perfectly well already how to enact command planning tasks and any system aimed at improving that should not require much, if anything, beyond which personnel bring to the task already. In summary, it seems wholly inappropriate to change the staff to fit the competencies required by the digital MP/BM system. Rather, the system should be engineered so that the tool set supports the current operators in the tasks that they are already considerably skilled at.

The Network in Network Enabled Capability

Changes to the current SOPs are expected; with new capabilities come new opportunities and new working processes. The development of the SOPs should be a symbiotic relationship with system development. As it is, the system under investigation is a peculiar hybrid, with the step-change represented by the digital system appended to a set of evolved SOPs. The high-speed hierarchy that results from this mix highlights that the 'Net' in NEC does not just refer to Networked technology, but also the type of organisation in which it is embedded.

Designing Digital MP/BM for Human Use

The proximal issue in terms of designing digital MP/BM for human use centres on the information presented by the system to its range of users. In every analysis performed on the system it is this issue that recurred; the design of the interface stunts human adaptability, represents a handicapping bottleneck for operational effectiveness, decreases tempo and increases complexity.

Firstly, it is clear that designers need to have a deep understanding of the process that they are designing to support; they need to understand the system at a number of levels of abstraction, its functional purpose, how this can be measured, the functions required and processes that need to take place. Further, they need to understand the relationships between these levels, particularly the second-order effects of changes to particular process. Designers also need to understand what the SA requirements of the various different end users are and design the system accordingly. In collaborative systems SA is an individual, team and systemic phenomenon and systems should support this by presenting global and specific user SA information. Collaborative systems should contain tools and interfaces tailored to the different SA requirements of its users. Schneiderman (1998) points out that, 'successful designers go beyond the vague notion of "user friendliness" probing deeper than simply making a checklist of subjective guidelines. They have a thorough understanding of the diverse community of users and the tasks that must be accomplished' (p. 10).

Secondly, as obvious as it sounds, the information presented by the system should be accurate at all times. It is surprising to observe, in the military domain of all domains, a system that does not present completely accurate information. Thirdly, timeliness of information presentation is key; the information presented by the system should be as up-to-date as is possible; delayed information diminishes user SA considerably.

The following guidance emerges from the Distributed Situation Awareness analysis performed:

1. *Clear definition and specification of SA (or information) requirements.* The findings suggest that the collaborative system design process should begin with a clear definition and specification of the SA requirements of the different users of the system in question. This should include a description of the process involved, the different roles and tasks involved in the process and a description of who needs to know what and when in the process they need to know it. Clearly, the designers of the system in this case did not fully appreciate the distinct roles and SA requirements of the different end users. Although this principle sounds somewhat obvious, unfortunately it is not always adhered to. Matthews et al. (2004) point out that knowing what the SA requirements are for a given domain provides engineers and technology developers a basis to develop optimal system designs

to maximise human performance rather than overloading workers and degrading their performance. Matthews et al. (2004) suggest that 'it is important, therefore, to know the SA requirements for various jobs to design systems that optimally present information, to evaluate the impact of new technology, and to develop effective training procedures to prepare workers to interact with advanced information systems' (p. 160). Matthews et al. also suggest that 'systematically identifying what it is the worker needs to know to accomplish key goals is a fundamental step in designing technological systems that optimise work performance' (p. 161) and that SA requirements analyses findings can be used to develop appropriate measures of SA for assessing the final system in terms of its support for SA requirements.

2. *Design system to support compatible SA requirements.* The findings suggest collaborative systems should be designed to cater for the compatible SA requirements of its end users. Within collaborative systems, users more often than not have distinct SA requirements and so the system should be designed so that users are not presented with information, tools and functionality that they do not explicitly require. The system should therefore be designed to support the roles, goals and SA requirements of each of the different users involved in the process. This might involve the provision of different displays, tools and functions for the varying roles and tasks involved. This removes the problem of high workload and getting bogged down in too much data and also reduces the requirement to send large products and data sets to every agent working within the system. In the same way that everyday PCs can be adapted by users so that the user interface and its functionality suit their own needs, it may also be more appropriate to allow the system and interface to be customisable based on the user's role (for example, G2) or on the job that the user is working on at a particular time (for example, synchronisation matrix) which will remove the vast number of redundant components of the system that get in the way when the user is doing their specific job. Gorman et al. (2006) advocate adaptive and timely information sharing, which they stress does not mean that everybody has access to the same information at the same time, but rather implies communicating the appropriate information (and importantly no more than this) to the right person at the right time. In this case the analysis indicated that the distinct SA requirements of the different end users were not supported in any way; rather the system remained the same in terms of information presentation, interfaces, tools available and functionality regardless of who was using it. The principle of providing system elements only with the information that they require becomes even more critical with the advent of NEC systems, where the great increases in information communicated around the system mean there is great potential for informational or data overload.

3. *Use multiple interlinked systems for multiple roles and goals.* When a team is divided into distinct roles and team members have very different goals and informational requirements it may be pertinent to offer separate (but linked) systems. In the same way that Microsoft Office provides separate word processing (for example, Word), drawing (for example, Visio) and spreadsheet (for example, Excel) tools, distributed team working support systems should provide a suite of mission support tools catering for the different users and roles involved; each tool should have the functionality and information required for the role it is designed to support whilst also containing the ability to see global information. As described previously the digital MP/BM system focussed on in the current study remained the same regardless of who was using it.

4. *Customisable/tailored interfaces.* As articulated previously, the nature of collaborative systems is such that there are specific roles and SA requirements. Subsequently, the

information and the tools that one agent needs to use may be very different to that that another agent needs. Collaborative systems should therefore be customisable, allowing users to customise (either by them or intelligently by the system based on usage) the interface so that the information and tools that they specifically require are present. This increases the usability and ease of use of the system and also reduces interaction time (that is, having to mine through menus to find information and tools required).

5. *Consider technological capability and impact on Distributed Situation Awareness.* Again perhaps an obvious, but nevertheless critical, recommendation is that system designers need to carefully consider the constraints imposed on them by technological capability and design the system accordingly. Distributed Situation Awareness in this analysis was adversely impacted by both the capability of the displays and mapping used and also by bandwidth limitations. It is therefore recommended that systems be designed within the constraints of the technology available.

6. *Ensure the accuracy of information presentation.* It goes without saying that the information presented by any command and control system should be highly accurate. System designers need to ensure that the information presented by all aspects of the system is accurate at all times. The present study revealed that the mission support system under analysis did not always present accurate SA-related information, such as contact and positional reports and enemy and friendly movements on the battlefield; further this information was often not presented in a timely manner.

7. *Provide filtering functions.* When systems have displays containing movement and location information relating to distinct entities (for example, enemy, friendly, neutral and so on) on a map, it is important that the system allows the users to filter the display so that different classes of information only are displayed.

8. *Clear communications links.* Throughout our research the importance of communications links for DSA acquisition and maintenance has consistently been highlighted; additionally a number of other researchers have identified communication links as key to team SA (for example, Gorman et al., 2006; Stanton et al., 2006; Walker et al., 2006) It is therefore critical that collaborative systems possess the appropriate communications links and that the users working with the system understand which communications channels are and are not open to them and also understand when and to whom what information should be communicated. This follows on from Stanton et al.'s (2006) conclusion that the links between agents in a network are at least as important as the agents themselves in maintaining DSA.

9. *Test DSA throughout the design lifecycle.* It is clear that DSA should be considered and tested where possible throughout the design life cycle. DSA requirements should be used to drive the design of concepts, and concepts should be evaluated based on their ability to meet the DSA requirements of the end users.

In conclusion, a clear paradox has emerged. At a more conceptual level the interface that has been designed is one that is extremely complex, one that forces users to perform a large number of relatively arbitrary, simplistic tasks. The paradox is that this is the polar opposite of the capability that it should afford. As highlighted within the analysis, digitisation is more than networked computers, it is about endowing command and control with synergistic, open-systems behaviour like 'self-synchronisation'. For this, a flexible, adaptable, simple system (from the user's perspective) is the sort of system that will allow humans to perform the complex military tasks for which they are selected, recruited and well able to perform.

Conclusions for Human Factors in Mission Planning

Moves have been made to develop digital systems to support distributed team working and planning processes (see Riley et al., 2006 and Roth et al., 2006) with the system that forms the topic of this book being the latest in a long line. The current system, like many of the previous, shares a focus on the products of the planning process for distribution between the planning team and to other people in the network. The challenge to system designers has been to preserve the collaborative, public and creative parts of the planning process as well as supporting different levels of plan fidelity (which will depend on the time available to develop the plan). Perhaps the biggest challenge is to decide what needs to be digitised and what form this digitisation should take. Given that military planning teams have invested considerable effort in developing and refining their planning skills using the traditional media, it would seem appropriate to try and support these activities, where appropriate, rather than requiring them to develop a new set of skills. Hybrid systems, therefore, may prove to be a viable point on the developmental timeline.

The planning process has evolved over centuries of refinement and improvement (Clausewitz, 1832). Roth et al. argue that much insight may be gleaned from studying the work-arounds and home-grown cognitive artefacts that are being used by command and control teams (such as the so-called 'cheat-sheets' and sticky notes). The traditional analogue planning process (as described earlier) is certainly abundant with potential metaphors, such as overlays, stickies, routes, Course of Actions (CoAs) and so on. It is worth considering if the conventional media could be captured digitally (by camera, scanner or other means) if they need to be transmitted as electronic documents with orders or reports, or for wider distribution. As a general design principle, the production of electronic documents should be at least as easy as the production of their analogue equivalents. Baxter (2005) is wary of the inexorable trend to digitise and concerned by the history of technology failing to deliver expected benefits, this is not just linked to military experience (Stanton & Marsden, 1996; Sinclair, 2007). Baxter argues that very few people understand the interrelated issues for technology, operations and Human Factors (being conversant in just one of these topics is not sufficient).

Transformational approaches of the 'step-change' nature seen throughout this book are likely to cause more problems than they solve. There are concerns that digitisation will lead to additional 'emergent' work (Kuper & Giurelli, 2007), both in terms of increasing the amount of 'direct' work required as well as the work associated with operation of the digital tools. The emergent nature of the task-artefact cycle has been described by Carroll (2000). Certainly it will not be possible to predict all the ways in which any future system would be used, so it is important to make the system as flexible as possible so that users may adapt and evolve it to suit their purposes (Roth et al., 2006).

Kiewiet et al. (2005) noticed that there are marked differences in the planners' domain knowledge, pointing out that group planning ensures an integrated approach rather than an overemphasis on one planner's area of strength. The social aspect of planning has not been lost on other researchers (Houghton et al., 2006; Stanton et al., 2006; Walker et al. 2006; Jenkins et al., 2008a). The collaborative aspects of planning seem to be a key to successful mission planning. As in the observational case study reported in this book, Riley et al. (2006) identified how different cells contributed to the planning process, such as intelligence, operations, logistics, fire support, engineering and air defence. Kuper & Giurelli (2007) argue that design of collaborative tools to support command and control teams is one of the keys to effective team work.

The case study presented by Riley et al. (2006) shows how Human Factors can contribute to the design of a mission planning system which is based on a thorough understanding of the planning process, the demands and constraints. In the design of their prototype tools they stress the need to provide a quick visualisation of the plan and the current situation. This enables the current operational picture to be compared with the plans, which may require changes to the plan as the situation changes (Stanton et al., 2008a; Stanton et al., 2008b). The Human Factors and Ergonomics methods introduced at the beginning of this book will help the individual to understand the demands and constraints placed on the people and technology in pursuit of their work, and therefore help design systems that are more appropriate. As a finale to the book we have imagined two futures for digital command and control, one pessimistic and one optimistic, which are presented as fictional 'thought experiments' based on the scenario described in chapter three (the mission planning process at BG).

In the first thought experiment, a pessimistic vision of technology is imagined. The WO has arrived from Bde, but this is a lengthy textual document that cannot be read easily off the small laptop screens. Printing off this large document is far from easy as it has a number of unintuitive steps. It is easy to forget to print off the attachments, and this process usually takes several attempts to complete. In addition, the printing is slow and at least 10 copies have to be printed so that all the staff officers have a copy. Once the WO is printed, the Chief of Staff (CoS) can go about the business of preparing the WOs for each company they are controlling. The CoS uses a different marker pen for each company and another to identify generic features in the orders. They then pass the order to a clerk who retypes the relevant material for each company into a word processor. The attachments are sorted from the Bde WO into a file on the main hard drive. As the file names are abstract, a separate list is kept on paper with a description of what is in each file. Sometimes this list is lost, so the files are required to be checked again and another paper log created. Occasionally the wrong attachment is sent to a company with their WOs. When the main orders arrive, the printing procedure is entered into again with similar frustrations, then a BG orders preparation timeline is created in the software. Unfortunately, the timeline is often longer than the screen, so the user has to keep scrolling left and right to see all of the deadlines for each of the Combat Estimate questions. The attachments cannot be used in their digital form, (for example, digital overlays for preparation of orders) partly because of incompatibilities in the graphics software (that is, there are problems with scaling of images as they appear in the wrong places) and partly because of the level of detail is not appropriate. Rather they have to print them off and then re-enter the information for themselves when recreating their own overlays. The planning timeline demands that the orders are produced within 2 hours, to allow the lower echelons time for their own planning. This means that there is 17 minutes per question in the Combat Estimates seven questions planning process. The drawing tools in the bespoke software do not conform to the normal industry standards (the ones that everyone is familiar with) and so the staff have to re-learn the functions of the tools every time they use them. In addition the user input devices are not as simple as a mouse (being a very stiff joystick that is 'soldier-proof'), which requires considerable force to move and the on-screen pointer moves jerkily. The staff officers complain that it is hard to find objects in the interface, and files are easily erased if the correct key combinations are not used when saving a file. This sometimes means that the same work is undertaken two or three times. The consequence of which is that it takes a considerable amount of time to produce the outputs for each part of the planning process. This is very frustrating for all involved. The software experts find they are being continually called to help with what they consider trivial aspects of the user interface, finding tools and icons, showing people how to print documents and attachments and reminding users how to save files. They do notice themselves, that after any time away from the software, they cannot remember how to perform some very basic functions, but they put this down to training fade and resolve to go on the next refresher course. The most frustrating aspect of the digital way of working for most users are all the additional tasks they have to do and that somehow an intuitive planning process has been turned into an unintuitive one in the name of 'progress'. Most frustrating of all is the fact that the planning process has been lengthened considerably. The 2-hour timeline to answer the seven questions now looks wildly optimistic, as they only got half way through by that time. But as one of the staff commented, that would have been plenty of time for the old analogue process, when paper maps, overlays, flip charts and pens were used for planning.

In the second thought experiment, an optimistic vision of technology is imagined. Again, the WO has arrived from Bde, this time the order is broken into chunks that can easily be read on the large, high definition, screens. To call it a document is a hang-up from the old world of paper. In fact it is a hypermedia webpage, with more in common with an Internet page. Clicking on links takes the reader to different parts of the orders, from the mission, to the maps, to the boundaries, to the assets, to the enemy. It is easy to forward on the relevant parts of the order, by a simple checking and highlighting process. Editing with notes and amendments to make the material relevant at company level is as simple as would have been with paper and pen for both the text and graphics pages. These pages are then 'posted' to each company by dragging and dropping into a mailbox and clicking on a send button. When the main orders arrive, everyone can see the orders webpage and click on the links relevant to their specialism. The large, high definition, screens make it easy to see the terrain in great amount of detail. As the screen can be laid flat or upright at the press of a button, it is easy for everyone to gather around and see the lay of the land. The relevant overlays from Bde can be edited by adding or removing symbols before they are overlaid on to the terrain maps. As it is a large touch screen, grabbing a symbol and positioning it is reminiscent of the old days of 'stickies' and 'acetates' – but these days the ground coordinates and details can appear automatically. Digital pens and rubbers mean that phase lines and boundaries are easy to add and remove, with all the data being stored automatically. As before, a timeline is created when the orders arrive, but the large screens mean that the timeline can be displayed in full. The drawing tools behave just like the tools in all other software so there is nothing new to learn. In fact, it is such an intuitive interface that most people don't comment on it as it's just a natural way of undertaking the work. Occasionally someone will notice that it is exceptionally easy to use (normally a visitor to the HQ) which will elicit a comment about how it is easier than the paper-based equivalent. The consequence of all this is that the timeline to produce the products is easily achieved with time to spare. The trainers of the system remark that there is no real training to be done, it's just plug-and-play command and control. Occasionally the old-timers will remark that the younger staff have it easy, as they had to learn the planning process the hard way with paper maps, overlays, flip charts and pens. Then, if you wanted multiple copies you had to redraw everything by hand, now you just press a button marked 'copy'. It's as simple as that.

It is worth pointing out that both of these thought experiments are entirely a figment of our collective imaginations, but it is our contention that Human Factors and Ergonomics can prevent the pessimistic outcome and assist in the optimistic one. It has been the aim of this book to show that Human Factors and Ergonomics has a considerable part to play in designing and evaluating command and control systems, be they digital or analogue.

References

Alberts, D. S. (2007) Agility, focus, and convergence: the future of command and control. *International C2 Journal*, 1(1), pp. 1–30.

Alberts, D. S. & Hayes, R. E. (2003) *Power to the Edge*. CCRP: Washington, DC.

Alberts, D. S. & Hayes, R. E. (2006) *Understanding Command and Control*. CCRP: Washington, DC.

Annett, J (2004) Hierarchical task analysis. In: D. Diaper and N. A. Stanton (Eds) *The Handbook of Task Analysis for Human Computer Interaction*. Lawrence Earlbaum Associates: Mahwah, NJ, pp. 67–82.

Annett, J., Duncan, K. D., Stammers, R. B. & Gray, M. J. (1971) Task analysis. Department of Employment Training Information Paper 6. HMSO, London.

Artman, H., & Garbis, C. (1998) Situation awareness as distributed cognition. In T. Green, L. Bannon, C. Warren & Buckley (Eds) Cognition and Cooperation. *Proceedings of 9th Conference of Cognitive Ergonomics*, (pp. 151–156). Limerick, Ireland.

Ashby, W. R. (1956) *Introduction to Cybernetics*. Chapman & Hall: London.

Baber, C. & Stanton, N. A. (1996) Human error identification techniques applied to public technology: Predictions compared with observed use. *Applied Ergonomics*, 27(2), pp. 119–131.

Bar Yam, Y. (2004) *Making Things Work: Solving Complex Problems in a Complex World*. NECSI Knowledge Press: Cambridge, MA.

Bartlett, F.C. (1932) *Remembering: A Study in Experimental and Social Psychology*. Cambridge University Press: Cambridge.

Baxter, R. (2005) Ned Ludd encounters Network Enabled Capability. *RUSI Defence Systems*, (Spring 2005), pp. 34–36.

Bevelas, A. (1948) A mathematical model for group structure. *Applied Anthropology*, 7, pp. 16–30.

Bolia, R., Vidulich, M, Nelson, T. & Cook, M. (2007) A history lesson in the use of technology to support military decision making and command and control. In M. Cook, J. Noyes and Y. Masakowski (Eds) *Decision Making in Complex Environments*. Ashgate: Aldershot, UK, pp. 191–200.

Bolstad, C. A., Riley, J. M., Jones, D. G., & Endsley, M. R. (2002) Using goal directed task analysis with Army brigade officer teams. In *Proceedings of the 46th Annual Meeting of the Human Factors and Ergonomics Society*, HFES: Baltimore, MD, pp. 472–476.

Bowers, C. A., Jentsch, J., Salas, E. & Braun, C. C. (1998) Analyzing the communication sequences for team training needs assessment. *Human Factors*, 40(4), pp. 672–679.

British Defence Film Library (BDFL) (2001) *Armoured Battle Group in the Quick Attack* # C003/00. Movie.

BS EN ISO 11064-6 (2005). *Ergonomic design of control centres – part 6: Environmental requirements for control centres*. BSI: London.

BS EN ISO 7730 (1995). *Moderate thermal environments – Determination of the PMV and PPD indices and specification of the conditions for thermal comfort*. BSI: London.

Burns, C. M., Bisantz, A. M. & Roth, E. M. (2004) Lessons from a comparison of work domain models: representational choices and their implications. *Human Factors*, 46 (4), pp. 711–727.

Burns, C. M., Bryant, D. J. & Chalmers, B. A. (2000) A work domain model to support shipboard command and control, in *International Conference on Systems, Man, and Cybernetics*, 2000, IEEE, Nashville, TN, USA, pp. 2228–2233.

Carroll, J. M. (2000) *Making Use: Scenario-Based Design of Human-Computer Interactions*. MIT Press: Mass.

CAST (2007), *The Digital Combat Estimate*. Command and Staff Trainer: Land Warfare Centre, Warminster (Unpublished MoD document).

Chapanis, A. (1999) *The Chapanis Chronicles*. Aegean Publishing Company: Santa Barbara, CA.

Chin, M., Sanderson, P. & Watson, M. (1999) Cognitive Work Analysis of the command and control work domain. *Proceedings of the 1999 Command and Control Research and Technology Symposium*, United States Naval War College: Newport, RI.

Clausewitz, C. Von (1832) On War (found at: http://www.clausewitz.com/CWZHOME/VomKriege2/ONWARTOC2.HTML accessed on: 15 August 2007).

Clegg, C. W. (2000) Sociotechnical principles for system design. *Applied Ergonomics*, 31, pp. 463–477.

Crone, D. J., Sanderson, P. M. & Naikar, N. (2003) Using Cognitive Work Analysis to Develop a Capability for the Evaluation of Future Systems. *Proceedings of the 47th Annual Meeting of the Human Factors and Ergonomics Society*, HFES: Denver, CO. pp. 1938–1942.

Crone, D., Sanderson, P., Naikar, N. & Parker, S. (2007) Selecting Sensitive Measures of Performance in Complex Multivariable Environments. *Proceedings of the 2007 Simulation Technology Conference* (SimTecT 2007). Brisbane, Australia, 4–7 June.

Crundell, B., Klein, G. and Hoffman, R. R. (2006) *Working Minds: A Practitioner's Guide to Cognitive Task Analysis*. MIT Press: Boston.

Cuevas, H. M., Costello, A. M., Bolstad, C. A. & Endsley, M. R. (2006) Facilitating Distributed Team Collaboration. *Proceedings of the International Ergonomics Association (IEA) 16th World Congress on Ergonomics*, Maastricht, The Netherlands, July 10–14.

Cummings, M. L. & Guerlain, S. (2003) The Tactical Tomahawk Conundrum: Designing Decision Support Systems for Revolutionary Domains, *IEEE Systems, Man, and Cybernetics Society conference*, Washington DC, October 2003.

Dekker, A. (2001) A category theoretic approach to social network analysis. *Electronic Notes in Theoretical Computer Science*, 61, pp. 1–13.

Dekker, A. (2002) Applying social network analysis concepts to military C4ISR architectures. *Connections*, 24(3), pp. 93–103.

Driskell, J. E. & Mullen, B. (2005) Social Network Analysis. In: N. A. Stanton, A. Hedge, K. Brookhuis, E. Salas, & H. Hendrick. (Eds), *Handbook of Human Factors and Ergonomics Methods*. (pp. 58.1–58.6) CRC: London.

Eklund, N. & Boyce, P. (1996) The development of a reliable, valid and simple office lighting survey. *Journal of the Illuminating Engineering Society*, 25, pp. 25–40.

Embrey, D. E. (1986). SHERPA: A systematic human error reduction and prediction approach. Paper presented at the *International Meeting on Advances in Nuclear Power Systems*, Knoxville, Tennessee.

Emery, F. E. & Trist, E. L. (1965) The causal texture of organizational environments. *Human Relations*, 18 (1), pp. 21–32.

Endsley, M. R. (1988) Situation Awareness Global Assessment Technique (SAGAT*). Proceedings of the National Aerospace and Electronics Conference (NAECON)*. (New York: IEEE), pp. 789–795.

Endsley, M. R. (1995a) Towards a theory of situation awareness in dynamic systems. *Human Factors*, 37, pp. 32–64.

Endsley, M. R. (1995b) Measurement of Situation Awareness in dynamic systems. *Human Factors*, 37, pp. 65–84.

Endsley, M. R. (1997) Situation Awareness: The Future of Aviation Systems. *Saab 60th Anniversary Symposium*, Linkoping, Sweden, Sept 8th.

Endsley, M. R. and Robertson, M. M. (2000) Situation Awareness in Aircraft Maintenance Teams. *International Journal of Industrial Ergonomics*, 26, 3001–325.

Endsley, M. R., & Jones, W. M. (1997) *Situation awareness, information dominance, and information warfare.* Technical Report 97-01. Endsley Consulting: Belmont, MA.

Endsley, M. R., Bolte, B., & Jones, D. G. (2003) *Designing for Situation Awareness: An Approach to User-centred Design.* Taylor & Francis: London.

Endsley, M. R., Holder, C. D., Leibricht, B.C., Garland, D.C., Wampler, R.L. & Matthews, M.D. (2000) *Modelling and Measuring Situation Awareness in the Infantry Operational Environment.* (1753). Army Research Institute: Alexandria, VA.

Engineering Equipment & Materials Users Association (EEMUA) (2001) *Process Plant Control Desks Utilising Human-Computer Interfaces – A Guide to Design Operational and Human Interface Issues*, EEMUA Publication 201. EEMUA: London.

Gleick, J. (1987) *Chaos: Making a New Science.* Cardinal: London.

Gorman, J. C., Cooke, N., & Winner, J. L. (2006) Measuring team situation awareness in decentralized command and control environments. *Ergonomics*, 49, pp. 1312–1326.

Grandjean, E. (1980) *Fitting the Task to the Man.* Taylor & Francis: London.

Grether, W. F. (1949) Instrument reading. 1. The design of long-scale indicators for speed and accuracy of quantitative readings. *Journal of Applied Psychology*, 33, pp. 363–372.

Hancock, P. A. (1997) *Essays on the Future of Human-Machine Systems.* Banta: Minneapolis.

Harary, F. (1994) *Graph Theory.* Addison-Wesley: Reading, MA.

Harris, C. J. & White, I. (1987) *Advances in Command, Control & Communication Systems.* Peregrinus: London.

Hedge, A. (2005) Indoor air quality: chemical exposures. In N. A. Stanton, A. Hedge, K. Brookhuis, E. Salas, & H. Hendrick. (Eds), *Handbook of Human Factors and Ergonomics Methods.* CRC Press: Boca Raton, FL.

Hedge, A. & Erickson, W. A. (1998) Indoor environment and sick building syndrome complaints in air conditioned offices: benchmarks for facility performance? *International Journal of Facilities Management*, 1, pp. 1–8.

Hollnagel, E. (1999) Cognitive systems engineering: new wine in new bottles. *International Journal of Human-Computer Studies*, 51, pp. 339–356.

Houghton, R. J., Baber, C., McMaster, R., Stanton, N. A., Salmon, P., Stewart, R. & Walker, G. H. (2006) Command and control in emergency services operations: a social network analysis. *Ergonomic*, 49(12–13), pp. 1204–1225.

HSE (2005). *Noise at work: Guidance for employers on the control of noise at work regulations 2005.* HMSO: London.

Hutchins, E. (1995) *Cognition in the Wild.* MIT Press: Cambridge MA.

ISO 9241-11 (1998) Ergonomic *requirements for office work with visual display terminals (VDTs) -- Part 11: Guidance on usability.* International Organization for Standardization, Geneva, Switzerland

Jenkins, D. P., Stanton, N. A., Salmon, P. M. & Walker, G. H (2009) *Cognitive Work Analysis: Coping with Complexity*. Ashgate: Farnham.

Jenkins, D. P., Stanton, N. A., Walker, G. H., Salmon, P. M. & Young, M. S. (2008a) Using Cognitive Work Analysis to explore activity allocation within military domains. *Ergonomics*, 51 (6), pp. 798–81

Jenkins, D. P., Stanton, N. A., Walker, G. H., Salmon, P. M. & Young, M. S. (2008b) Applying Cognitive Work Analysis to the design of rapidly reconfigurable interfaces in complex networks. Theoretical Issues in Ergonomics Science, 9 (4) pp. 273–295.

Jones, D. G., Bolstad, C. A., Riley, J. M. & Endsley M. R. (2003) Situation Awareness requirements for the future objective force. Paper presented at the *Collaborative Technology Alliances Symposium* April 29th–May 1st 2003 College Park, MD.

Kakimoto, T., Kamei, Y., Ohira, M. & Matsumoto, K. (2006) Social Network Analysis on communications for knowledge collaboration in OSS communities. In *Proc. The 2nd International Workshop on Supporting Knowledge Collaboration in Software Development* (KCSD'06), pp. 35–41. Tokyo, Japan.

Kiewiet, D. J., Jorna, R. J. and Wezel, W. V. (2005) Planners and their cognitive maps: an analysis of domain representation using multi-dimensional scaling. *Applied Ergonomics*, 36, pp. 695–708.

Kirwan, B. & Ainsworth, L. K. (1992) *A Guide to Task Analysis*. Taylor & Francis: London.

Klein, G. & Armstrong, A. A. (2004) Critical decision method. In N. A. Stanton, A. Hedge, E. Salas, H. Hendrick, & K. Brookhaus, (Eds), *Handbook of Human Factors and Ergonomics Methods* (pp. 35.1–35.8), CRC Press: Boca Raton, Florida.

Klein, G. & Miller, T. E. (1999) Distributed planning teams. *International Journal of Cognitive Ergonomics*, 3 (3), pp. 203–222.

Kuper, S. R. & Giurelli, B. L. (2007) Custom work aids for distributed command and control teams: a key to enabling highly effective teams. *The International C2 Journal*, 1 (2), pp. 25–42.

Lamoureux, T. M., Rehak, L. A., Bos, J. C. & Chalmers, B. (2006) Control Task Analysis in applied settings. *Proceedings of the Human Factors and Ergonomics Society 50th Annual Meeting 2006*, HFES: XXX pp. 391–395.

Lawson, J. S. (1981) Command and control as a process. *IEEE Control Systems Magazine*, March, pp. 86–93.

Leavitt, H. J. (1951) Some effects of certain communication patterns on group performance. *Journal of Abnormal and Social Psychology*, 46, pp. 38–50.

Lee, J. D. (2001) Emerging challenges in cognitive ergonomics: managing swarms of self-organising agent-based automation. *Theoretical Issues in Ergonomics Science*, 2 (3), pp. 238–250.

Levchuk, G. M., Levchuk, Y. N., Luo, J., Pattipati, K. R. & Kleinman, D. L. (2002) Normative design of organisations – Part I: Mission Planning. *IEEE Transactions on Systems, Man and Cybernetics – Part A: Systems and Humans*, 32 (3), pp. 346–359.

Lintern, G., Cone, S., Schenaker, M., Ehlert, J. & Hughes, T. (2004) *Asymmetric Adversary Analysis for Intelligent Preparation of the Battlespace* (A3-IPB) United States Air Force Research Department Report.

Macey, P. (2007a) *Scientific Task Terms of Reference for OFT3*, V0.2, 3rd Sept 2007 C2DC, Land Warfare Centre: Warminster.

Macey, P. (2007b) *Critical Success Factors*. C2DC internal report. C2DC, Land Warfare Centre, Warminster.

Matthews, M. D., Strater, L. D. & Endsley, M. R. (2004) Situation awareness requirements for infantry platoon leaders. *Military Psychology*, 16 (3), pp 149–161.

Mauer, M. E. (1994) *Coalition Command and Control*. Washington National Defense University.

Meister, D. (1989) *Conceptual Aspects of Human Factors*. John Hopkins University Press: Baltimore.

Meister, D. (1999) *The History of Human Factors and Ergonomics*. Lawrence Erlbaum Associates: Mahwah, NJ.

Ministry of Defence (2000) *Human Factors Integration: An introductory guide*. HMSO: London.

Moltke, H. K. B. G. von (undated) http://en.wikipedia.org/wiki/Helmuth_von_Moltke_the_Elder (accessed on 15 August 2007).

Murrell, K. F. H. (1965) *Human Performance in Industry*. Reinhold Publishing: New York.

Naikar, N. & Sanderson, P.M. (2001) Evaluating design proposals for complex systems with work domain analysis. *Human Factors*, 43, pp. 529–542.

Naikar, N. & Saunders, A. (2003) Crossing the boundaries of safe operation: A technical training approach to error management. *Cognition Technology and Work*, 5, pp. 171–180.

NATO (2006) SAS-050 Exploring new command and control concepts and capabilities: Final report. Available at: http://www.dodccrp.org/SAS/SAS-050%20Final%20Report.pdf.

Neisser, U. (1976) *Cognition and Reality: Principles and Implications of Cognitive Psychology*. Freeman: San Francisco.

Norman, D. (2007a) *The Design of Future Things*, Basic Books (Perseus): New York.

Norman, D. (2007b) UI Breakthrough-Command Line Interfaces, on-line article available at http://www.jnd.org/dn.mss/ui_breakthroughcomma.html

Oborne, D. J. (1982) *Ergonomics at Work*. Wiley: Chichester.

Prefontaine, R. (2002) The administrative estimate in the operation planning process: a toll not well understood. *The Army Doctrine and Training System*, 5 (4), pp. 4–12.

Rasmussen, J. (1985). The role of hierarchical knowledge representation in decision making and system management. *IEEE Transactions on Systems, Man and Cybernetics*, 15, pp. 234–243

Ravden, S. J. and Johnson, G. I. (1989) *Evaluating Usability of Human-Computer Interfaces: A Practical Method*. Ellis Horwood: Chichester.

Reason, J. (1997) *Managing the Risks of Organisational Accidents*. Burlington, VT: Ashgate

Riley, J. M., Endsley, M. R., Bolstad, C. A., & Cuevas, H. M. (2006) Collaborative planning and situation awareness in army command and control. *Ergonomics*, 49, pp. 1139–1153.

Rittel, H. & Webber, M. (1973) Dilemmas in a General Theory of Planning, *Policy Sciences*, 4, pp. 155–169 [Reprinted in N. Cross (Ed.), (1984) *Developments in Design Methodology*, J. Wiley & Sons: Chichester, pp. 135–144].

Roth, E., Scott, R., Deutsch, S., Kuper, S., Schmidt, V., Stilson, M. & Wampler, J. (2006) Evolvable work-centred support systems for command and control. *Ergonomics*, 49 (7), pp. 688–705.

Salmon, P. M, Stanton, N. A., Baber, C., Walker, G. H., McMaster, R. & Jenkins, D. P. (2008) Representing Situation Awareness in collaborative systems: A case study in the energy distribution domain. *Ergonomics*, 51 (3), pp. 367–384.

Salmon, P.M., Stanton, N. A., Walker, G., & Green, D. (2006) Situation Awareness measurement: A review of applicability for C4i environments. *Applied Ergonomics*, 37, pp. 225–238.

Salmon, P. M., Stanton, N. A., Walker, G. H., & Jenkins, D. P. (2009) *Distributed Situation Awareness: Advances in Theory, Measurement and Application to Teamwork*. Ashgate: Aldershot, UK.

Sanders, M. S. & McCormick, E. J. (1993) *Human Factors in Engineering and Design.* McGraw-Hill: New York.

Schneiderman, B. (1998) *Designing the User Interface: Strategies for Effective Human-Computer Interaction.* 3rd Edition, Addison Wesley Longmann, Inc: USA.

Shachtman, N. (2007) How technology almost lost the war: In Iraq, the critical networks are social – not electronic. *Wired Magazine,* 15, (12).

Shadbolt, N. R. & Burton, M. (1995) Knowledge elicitation: A systemic approach. In J. R. Wilson and E. N. Corlett (Eds) *Evaluation of Human Work: A Practical Ergonomics Methodology,* pp. 406–440.

Shepherd, A. (1985) Hierarchical task analysis and training decisions. *Programmed Learning and Educational Technology,* 22, pp. 162–176.

Shu, Y. & Furuta, K. (2005) An inference method of team situation awareness based on mutual awareness. *Cognition Technology & Work,* 7, pp. 272–287.

Sinclair, M. A. (2007) Ergonomics issues in future systems. *Ergonomics,* 50 (12), pp. 1957–1986.

Stanton, N. A. (2006) Hierarchical task analysis: Developments, applications and extensions. *Applied Ergonomics,* 37, pp. 55–79.

Stanton, N. A. & Marsden, P. (1996) From fly-by-wire to drive-by-wire: safety implications of automation in vehicles. *Safety Science,* 24 (1) pp. 35–49.

Stanton, N. A., & Young, M. S. (1999) *A Guide to Methodology in Ergonomics: Designing for Human Use.* Taylor and Francis: London.

Stanton, N. A., Baber, C. & Harris, D. (2008a) *Modelling Command and Control: Event Analysis of Systemic Teamwork.* Ashgate: Aldershot.

Stanton, N.A., Baber, C., Walker, G.H., Houghton, R.J., McMaster, R., Stewart, R., Harris, D., Jenkins, D.P., Young, M.S. & Salmon, P.M. (2008) Development of a generic activities model of command and control. *Cognition, Technology and Work,* 10 (3) pp. 209–220.

Stanton, N. A., Hedge, A., Brookhuis, K., Salas, E. & Hendrick, H. (2005b) *Handbook of Human Factors and Ergonomics Methods.* CRC Press: Boca Raton, Fl.

Stanton, N. A., Salmon, P. M. & Walker, G. H. (2009) Genotype and phenotype schemata as models of situation awareness in dynamic command and control teams. *International Journal of Industrial Ergonomics,* 39(3), pp. 480–489.

Stanton, N. A., Salmon, P. M., Walker, G. H., Baber, C. & Jenkins, D. P. (2005a) *Human Factors Methods: A Practical Guide for Engineering and Design.* Ashgate: Aldershot.

Stanton, N. A., Stewart, R., Harris, D., Houghton, R. J., Baber, C., McMaster, R., Salmon, P. M., Hoyle, G., Walker, G. H., Young, M. S., Linsell, M., Dymott, R. & Green, D. (2006) Distributed situation awareness in dynamic systems: theoretical development and application of an ergonomics methodology. *Ergonomics,* 49, pp. 1288–1311.

Stewart, R., Stanton, N. A., Harris, D., Baber, C., Salmon, P.M., Mock, M., Tatlock, K., Wells, L. & Kay, A. (2008) Distributed situation awareness in an airborne warning and control system: application of novel ergonomics methodology. *Cognition Technology and Work,* 10, 3, pp. 221–229.

Taylor, R. M. (1990) Situational awareness rating technique (SART): The development of a tool for aircrew systems design. In *Situational Awareness in Aerospace Operations* (AGARD-CP-478) pp. 3/1–3/17, Neuilly Sur Seine, France: NATO-AGARD.

Torres, R. R. (2005) Rapid sound-quality assessment of background noise. In N. A. Stanton, A. Hedge, K. Brookhuis, E. Salas, & H. Hendrick. (Eds), *Handbook of Human Factors and Ergonomics Methods.* CRC Press: Boca Raton, FL.

Vicente, K. J. (1999) *Cognitive Work Analysis: Toward Safe, Productive, and Healthy Computer-based Work*. Lawrence Erlbaum Associates: Mahwah, NJ.

Waldrop, M. M. (1992) *Complexity: The Emerging Science at the Edge of Order and Chaos*. Simon & Schuster: New York.

Walker, G. H., Stanton, N. A., Salmon, P. M. & Jenkins, D. P. (2009). A review of sociotechnical systems theory: A classic concept for new command and control paradigms. *Theoretical Issues in Ergonomics Science*, 9(6), pp. 479–499.

Walker, G. H., Stanton, N. A., Salmon, P., Jenkins, D., Stewart, R. & Wells, L. (2009) Using an integrated methods approach to analyse the emergent properties of military command and control. *Applied Ergonomics*, 40(4), pp. 636–647.

Walker, G.H., Gibson, H., Stanton, N.A., Baber, C., Salmon, P. & Green, D. (2006) Event analysis of systemic teamwork (EAST): A novel integration of ergonomics methods to analyse C4i activity, *Ergonomics*, 49 (12–13), pp. 1345–1369.

Warm, J. S., Dember, W. N., & Hancock, P. A. (1996) Vigilance and workload in automated systems. In R. Parasuraman and M. Mouloua (Eds), *Automation and Human Performance: Theory and Applications*. (pp. 183–200). Lawrence Erlbaum Associates: Mahwah, NJ.

Weisstein, E W. (accessed on 17th Sept 2008) Graph Diameter. From MathWorld--A Wolfram Web Resource. Available at: http://mathworld.wolfram.com/GraphDiameter.html.

Woods, D. D. (1988) Coping with complexity: The psychology of human behaviour in complex systems. In L.P. Goodstein, H.B. Andersen and S.E. Olsen (Eds), *Tasks, Errors and Mental Models*, Taylor Francis: London, pp. 128–148.

Index

Author Index